Ben Stacy Jerrik (Ed.)

# Hüniken

Ben Stacy Jerrik (Ed.)

# Hüniken

## Wasseramt District, Cantons of Switzerland, Communes of Switzerland

Part Press

**Imprint**



Publisher:
Part Press is a trademark of
International Book Market Service Ltd., 17 Rue Meldrum, Beau Bassin, 1713-01 Mauritius
Email: info@bookmarketservice.com
Website: www.bookmarketservice.com

Published in 2013

Printed in: U.S.A., U.K., Germany. This book was not produced in Mauritius.

**ISBN: 978-620-1-59250-6**

# Contents

## Articles



## References

# Richland,_Pennsylvania

**Richland** is a borough in Lebanon County, Pennsylvania, United States. It is part of the Lebanon, PA Metropolitan Statistical Area. The population was 1,508 at the 2000 census.

## Geography

Richland is located at (40.357122, -76.257187)[] .

According to the United States Census Bureau, the borough has a total area of 1.6 square miles ( $km^2$), all of it land.

## Demographics

As of the census[] of 2000, there were 1,508 people, 582 households, and 444 families residing in the borough. The population density was 965.7 people per square mile (373.2/km²). There were 602 housing units at an average density of 385.5 per square mile (149.0/km²). The racial makeup of the borough was 98.21% White, 0.33% African American, 0.33% Asian, 0.33% from other races, and 0.80% from two or more races. Hispanic or Latino of any race were 0.60% of the population.

There were 582 households out of which 32.8% had children under the age of 18 living with them, 66.0% were married couples living together, 6.0% had a female householder with no husband present, and 23.7% were non-families. 19.8% of all households were made up of individuals and 10.5% had someone living alone who was 65 years of age or older. The average household size was 2.59 and the average family size was 2.96.

In the borough the population was spread out with 25.1% under the age of 18, 6.6% from 18 to 24, 29.4% from 25 to 44, 22.5% from 45 to 64, and 16.4% who were 65 years of age or older. The median age was 38 years. For every 100 females there were 97.4 males. For every 100 females age 18 and over, there were 93.5 males.

The median income for a household in the borough was $45,729, and the median income for a family was $52,063. Males had a median income of $35,208 versus $22,723 for females. The per capita income for the borough was $19,365. About 2.3% of families and 3.2% of the population were below the poverty line, including 2.7% of those under age 18 and 6.0% of those age 65 or over.

## Fire Company

http://www.neptunefire.com/

## History

Richland is unusual in that an active railroad crossing intersects the town square. Further, the two streets that comprise the square, Main Street and Race Street, are the only streets linking the northern and southern portions of the borough. As a result, that crossing can divide the entire town. This quirk has earned Richland mention in *Ripley's Believe It Or Not* books and on the televised game show, *Jeopardy*.[1]

## References

[1] http://www.joellenlitz.com/7_wonders_of_lebanon_county.htm

# Borough

A **borough** is an administrative division in various countries. In principle, the term *borough* designates a self-governing township although, in practice, official use of the term varies widely.

The word *borough* derives from common Germanic **burg*, meaning *fort*: compare with *bury* (England), *burgh* (Scotland), *Burg* (Germany), *borg* (Scandinavia), *pori* (Finland), *burcht* (Dutch), and the Germanic borrowing present in neighbouring Indo-european languages such as *borgo* (Italian), *bourg* (French), *burgo* (Spanish and Portuguese), *purg* (Kajkavian) and *durg* (Hindi). The incidence of these words as suffixes to place names (for example, Canterbury, Strasbourg, Luxembourg, Edinburgh, Hamburg, Gothenburg) usually indicates that they were once fortified settlements.

In the Middle Ages, boroughs were settlements in England that were granted some self-government; burghs were the Scottish equivalent. In medieval England, boroughs were also entitled to elect members of parliament. The use of the word *borough* probably derives from the burghal system of Alfred the Great. Alfred set up a system of defensive strong points (Burhs); in order to maintain these settlements, he granted them a degree of autonomy. After the Norman Conquest, when certain towns were granted self-governance, the concept of the burh/borough seems to have been reused to mean a self-governing settlement.

The concept of the borough has been used repeatedly (and often differently) throughout the world. Often, a borough is a single town with its own local government. However, in some cities it is a subdivision of the city (for example, London, New York City, Montreal, and Tokyo). In such cases, the borough will normally have either limited powers delegated to it by the city's local government, or no powers at all. At certain times, London has had no overall city government and boroughs were the main unit of local government for Londoners; in Tokyo, they are known as wards or cities, and have city status, but these still form parts of a larger municipal government which has merged with that of the prefecture. In other places, such as Alaska, *borough* designates a whole region; Alaska's largest borough, the North Slope Borough, is comparable in area to the entire United Kingdom, although its population is less than that of Swanage. In Australia, *borough* can designate a town and its surrounding area, such as the Borough of Queenscliffe.

Boroughs as administrative units are to be found in Ireland and the United Kingdom, more specifically in England and Northern Ireland. Boroughs also exist in the Canadian province of Quebec and formerly in Ontario, in some states of the United States, in Israel, and formerly in New Zealand.

## Etymology

The word *borough* derives from the Old English word *burh*, meaning a fortified settlement. Other English derivatives of *burh* include *bury* and *brough*. There are obvious cognates in other Indo-European languages. For example; *burgh* in Scots and Middle English; *burg* in German and Old English,[1] *borg* in Scandinavian languages; *parcus* in Latin and *pyrgos* in Greek.

A number of other European languages have cognate words that were borrowed from the Germanic languages during the Middle Ages, including *brog* in Irish, *bwr* or *bwrc*, meaning "wall, rampart" in Welsh, *bourg* in French, *burg* in Catalan (in Catalonia there is a town named *Burg*), *borgo* in Italian, and *burgo* in Spanish (hence the place-name Burgos).

The 'burg' element is often confused with 'berg' meaning hill or mountain (cf. iceberg). Hence the 'berg' element in Bergen relates to a hill, rather than a fort. In some cases, the 'berg' element in place names has converged towards burg/borough; for instance Farnborough, from *fernaberga* (fern-hill).

## Pronunciation

In many parts of England, "borough" is pronounced 🔊 [i]/ˈbʌrə/ as an independent word, and as /brə/ when a suffix of a place-name. As a suffix, it is sometimes spelled "-brough".

In the United States, "borough" is pronounced /ˈbʌroʊ/ or /ˈbʊəroʊ/. When appearing as the suffix "-burg(h)" in place-names, it is pronounced /bɜrg/.

## Uses of *borough*

### England and Wales

#### Ancient and municipal boroughs

During the medieval period many towns were granted self-governance by the Crown, at which point they became referred to as boroughs. The formal status of borough came to be conferred by Royal Charter. These boroughs were generally governed by a self-selecting corporation (i.e., when a member died or resigned his replacement would be by co-option). Sometimes boroughs were governed by bailiffs or headboroughs.

Debates on the Reform Bill (eventually the Reform Act 1832) had highlighted the variations in systems of governance of towns, and a Royal Commission was set up to investigate the issue. This resulted in a regularisation of municipal government (Municipal Corporations Act 1835). 178 of the ancient boroughs were reformed as *municipal boroughs*, with all municipal corporations to be elected according to a standard franchise based on property ownership. The unreformed boroughs either lapsed in borough status, or were reformed (or abolished) at a later time. Several new municipal boroughs were formed in the new industrial cities after the bill enacted, according to the provisions of the bill.

As part of a large-scale reform of local government in England and Wales in 1974, municipal boroughs were finally abolished (having become increasingly irrelevant). However, the civic traditions of many boroughs were continued by the grant of a charter to their successor district councils. In smaller boroughs, a town council was formed for the area of the abolished borough, while charter trustees were formed in other former boroughs. In each case, the new body was allowed to use the regalia of the old corporation, and appoint ceremonial office holders such as sword and mace bearers as provided in their original charters. The council or trustees may apply for an Order in Council or Royal Licence to use the former borough coat of arms.

#### Parliamentary boroughs

Further information: Reform Act 1832

From 1265, two burgesses from each borough were summoned to the Parliament of England, alongside two knights from each county. Thus parliamentary constituencies were derived from the ancient boroughs. Representation in the House of Commons was decided by the House itself, which resulted in boroughs' being established in some small settlements for the purposes of parliamentary representation, despite their possessing no actual corporation.

After the Reform Act, which disenfranchised many of the rotten boroughs (boroughs that had declined in importance, had only a small population, and had only a handful of eligible voters), parliamentary constituencies began to diverge from the ancient boroughs. While many ancient boroughs remained as municipal boroughs, they were disenfranchised by the Reform Act.

### County boroughs

The Local Government Act 1888 established a new sort of borough – the county borough. These were designed to be 'counties-to-themselves'; administrative divisions to sit alongside the new administrative counties. They allowed urban areas to be administered separately from the more rural areas. They, therefore, often contained pre-existing municipal boroughs, which thereafter became part of the second tier of local government, below the administrative counties and county boroughs.

The county boroughs were, like the municipal boroughs, abolished in 1974, being reabsorbed into their parent counties for administrative purposes.

### Metropolitan boroughs

In 1899, as part of a reform of local government in the County of London, the various parishes in London were reorganised as new entities, the 'metropolitan boroughs'. These were reorganised further when Greater London was formed out of Middlesex and the County of London in 1965.

When the new metropolitan counties (Greater Manchester, Merseyside, Tyne & Wear, West Midlands, South Yorkshire and West Yorkshire) were created in 1974, their sub-divisions also became metropolitan boroughs; in many cases these metropolitan boroughs recapitulated abolished county boroughs (for example, Stockport). The metropolitan boroughs possessed slightly more autonomy from the metropolitan county councils than the shire county districts did from their county councils.

With the abolition of the metropolitan county councils in 1986, these metropolitan boroughs became independent, and continue to be so at present.

### Other current uses

Elsewhere in England a number of districts and unitary authority areas are called "borough". Until 1974, this was a status that denoted towns with a certain type of local government (a municipal corporation). Since 1974, it has been a purely ceremonial style granted by royal charter to districts which may consist of a single town or may include a number of towns or rural areas. Borough status entitles the council chairman to bear the title of mayor. Districts may apply to the British Crown for the grant of borough status upon advice of the Privy Council of the United Kingdom.

## Northern Ireland

In Northern Ireland, local government was reorganised in 1973. Under the legislation that created the 26 districts of Northern Ireland, a district council whose area included an existing municipal borough could resolve to adopt the charter of the old municipality and thus continue to enjoy borough status. Districts that do not contain a former borough can apply for a charter in a similar manner to English districts.

## Canada

In Quebec, the term **borough** is generally used as the English translation of arrondissement, referring to an administrative division of a municipality. Only eight municipalities in Quebec are divided into boroughs. See List of boroughs in Quebec.

It was previously used in Metropolitan Toronto, Ontario, to denote suburban municipalities including Scarborough, York, North York, Etobicoke prior to their conversion into cities. The Borough of East York was the last Toronto municipality to hold this status, relinquishing it upon becoming part of the City of Toronto on January 1, 1998.

## United States

In the United States, a borough is a unit of local government below the level of the state. The term is currently used in seven states.

The following states use, or have used, the word with the following meanings:

- Alaska, as a county-equivalent
- Connecticut, as an incorporated municipality within, or consolidated with, a town
- Minnesota, formerly applied to one municipality
- New Jersey, as a type of independent incorporated municipality - see Borough (New Jersey)
- New York, as one of the five divisions of New York City, each coextensive with a county - See Borough (New York City)
- Pennsylvania, as a type of municipality comparable to a town. Only one incorporated town is chartered in Pennsylvania. - see Borough (Pennsylvania)
- Virginia, as a division of a city under certain circumstances.

Certain names of places, such as Hillsboro, OR, Greensboro, NC, Tyngsborough, MA, and Maynesborough, NH reflect the historical use of Borough as a geographical scale in the United States.[2]

## Mexico

In Mexico as translations from English to Spanish applied to Mexico City, the word *borough* has resulted in a **delegación** (delegation), referring to the 16 administrative areas within the Mexican Federal District. Also the municipalities of some states are administratively subdivided into boroughs, as showed in Municipality of Mexicali#Boroughs. (**see**: Boroughs of Mexico and Boroughs of the Mexican Federal District)

## Australia

In Australia, the term "borough" is an occasionally used term for a local government area. Currently there is only one borough in Australia, the Borough of Queenscliffe in Victoria, although there have been more in the past. However in some cases it can be integrated into the councils name instead of used as an official title, such as the Municipality of Kingborough in Tasmania.

## Republic of Ireland

Under the Local Government Act 2001 section 10 (3) and schedule 6 part 1 chapter 1, the following continue to be known as Boroughs (though this is largely a matter of nomenclature) Clonmel, Drogheda, Kilkenny, Sligo, Wexford. In Section 10 (7) continues the "use of the description city in relation to Kilkenny, to the extent that that description was used before (January 1, 2002) and is not otherwise inconsistent with this Act."

## New Zealand

New Zealand formerly used the term borough to designate self-governing towns of more than 1,000 people, although 19th century census records show many boroughs with populations as low as 200.[3] A borough of more than 20,000 people could become a city by proclamation. Boroughs and cities were collectively known as municipalities, and were enclaves separate from their surrounding counties. Boroughs proliferated in the suburban areas of the larger cities: By the 1980s there were 19 boroughs and three cities in the area that is now the City of Auckland.

In the 1980s, some boroughs and cities began to be merged with their surrounding counties to form districts with a mixed urban and rural population. In 1989, a nationwide reform of local government completed the process. Counties and boroughs were abolished and all boundaries were redrawn. Under the new system, most territorial authorities cover both urban and rural land. The more populated councils are classified as cities, and the more rural councils are classified as districts. Only Kawerau District, an enclave within Whakatane District, continues to follow the tradition of a small town council that does not include surrounding rural area.

### Israel

Under Israeli law, inherited from British Mandate municipal law, the possibility of creating a municipal borough exists. However, no borough was actually created under law until 2005–2006, when Neve Monosson and Maccabim-Re'ut, both communal settlements (Heb: ***yishuv kehilati***) founded in 1953 and 1984, respectively, were declared to be autonomous municipal boroughs (Heb: ***vaad rova ironi***), within their mergers with the towns of Yehud and Modi'in. Similar structures have been created under different types of legal status over the years in Israel, notably Kiryat Haim in Haifa, Jaffa in Tel Aviv-Yafo and Ramot and Gilo in Jerusalem. However, Neve Monosson is the first example of a full municipal borough actually declared under law by the Minister of the Interior, under a model subsequently adopted in Maccabim-Re'ut as well.

It is the declared intention of the Interior Ministry to use the borough mechanism in order to facilitate municipal mergers in Israel, after a 2003 wide-reaching merger plan, which, in general, ignored the sensitivities of the communal settlements, and largely failed.

### Netherlands

Larger cities in the Netherlands are often divided into boroughs, or stadsdelen

## See also

- Borough status in the United Kingdom
- Boroughs incorporated in England and Wales 1835–1882
- Boroughs incorporated in England and Wales 1882–1974
- Boroughs in New York City
- Burgh
- County borough
- Ancient borough
- History of local government in England
- List of burghs in Scotland
- Metropolitan borough
- Municipal borough

## References

[1] The American Heritage Dictionary of the English Language: Fourth Edition (2000)
[2] Words Ending with Boro (http://www.whatrhymeswith.com/words-ending-with/boro)
[3] 1881 census summary (http://freepages.genealogy.rootsweb.ancestry.com/~sooty/nzcensus1881.html)

## External links

-

# Lebanon_County,_Pennsylvania

**Lebanon County** is a county located in the U.S. state of Pennsylvania. As of the 2010 census, the population was 133,568. Its county seat is Lebanon.[]

Lebanon County is part of the Lebanon Metropolitan Statistical Area and the Harrisburg–Carlisle–Lebanon Combined Statistical Area.

## History

Lebanon County was formed from portions of Dauphin and Lancaster counties in 1813, with minor boundary revisions in 1814 and 1821.[1]

## Geography

According to the U.S. Census Bureau, the county has a total area of 363 square miles ( $km^2$), of which 362 square miles ( $km^2$) is land and 1 square mile ( $km^2$) (0.20%) is water. It is drained by Swatara Creek and its branches. It consists in large part of a valley.[2]

### Adjacent counties

- Schuylkill County (northeast)
- Berks County (east)
- Lancaster County (south)
- Dauphin County (west){north}

## Demographics

| Historical populations | | |
|---|---|---|
| **Census** | **Pop.** | **%±** |
| **1820** | 16988 | — |
| **1830** | 20557 | 21.0% |
| **1840** | 21872 | 6.4% |
| **1850** | 26071 | 19.2% |
| **1860** | 31831 | 22.1% |
| **1870** | 34096 | 7.1% |
| **1880** | 38476 | 12.8% |
| **1890** | 48131 | 25.1% |
| **1900** | 53827 | 11.8% |
| **1910** | 59565 | 10.7% |
| **1920** | 63152 | 6.0% |
| **1930** | 67103 | 6.3% |
| **1940** | 72641 | 8.3% |
| **1950** | 81683 | 12.4% |
| **1960** | 90853 | 11.2% |

| | | |
|---|---|---|
| **1970** | 99665 | 9.7% |
| **1980** | 108582 | 8.9% |
| **1990** | 113744 | 4.8% |
| **2000** | 120327 | 5.8% |
| **2010** | 133568 | 11.0% |
| [3] [4] | | |

As of the census[] of 2000, there were 120,327 people and 32,771 families residing in the county. The population density was 332 people per square mile (128/km²). There were 49,320 housing units at an average density of 136 per square mile (53/km²). The racial makeup of the county was 94.46% White, 1.29% Black or African American, 0.13% Native American, 0.89% Asian, 0.04% Pacific Islander, 2.26% from other races, and 0.94% from two or more races. 4.96% of the population were Hispanic or Latino of any race. 45.6% were of German, 11.8% American and 6.1% Irish ancestry according to Census 2000. 92.5% spoke English, 4.2% Spanish and 1.1% Pennsylvania Dutch as their first language.

There were 46,551 households out of which 30.40% had children under the age of 18 living with them, 57.40% were married couples living together, 9.20% had a female householder with no husband present, and 29.60% were non-families. 25.20% of all households were made up of individuals and 11.10% had someone living alone who was 65 years of age or older. The average household size was 2.49 and the average family size was 2.98.

In the county, the population was spread out with 23.70% under the age of 18, 8.20% from 18 to 24, 28.00% from 25 to 44, 23.70% from 45 to 64, and 16.40% who were 65 years of age or older. The median age was 39 years. For every 100 females there were 95.00 males. For every 100 females age 18 and over, there were 91.70 males.

## Politics and Government

### United States House of Representatives

The county is divided into the 6th congressional district, represented by Jim Gerlach, Republican, and the 15th congressional district, represented by Charlie Dent, Republican.

### Pennsylvania State Senate

All of the county falls within the 48th Pennsylvania Senatorial District. The seat is currently held by State Sen. Mike Folmer, a Republican.

### Pennsylvania House of Representatives

The county is divided into the 101st and 102nd Pennsylvania House Districts.

#### 101st District

The 101st District, served by Republican Mauree Gingrich, whose term expires in 2013, includes:

- Annville Twp.
- Cleona Borough
- Lebanon City
- Mt. Gretna Borough
- North Cornwall Twp.
- North Londonderry Twp.
- Palmyra Borough
- South Annville Twp.
- South Londonderry Twp.

#### 102nd District

The 102nd District, served by Republican RoseMarie Swanger, whose term expires in 2013, includes:

- Bethel Twp.
- Cold Spring Twp.
- Cornwall Borough
- East Hanover Twp.
- Heidelberg Twp.
- Jackson Twp.
- Jonestown Borough
- Millcreek Twp.
- Myerstown Borough
- North Annville Twp.
- North Lebanon Twp.
- Richland Borough
- South Lebanon Twp.
- Swatara Twp.
- Union Twp.
- West Cornwall Twp.
- West Lebanon Twp.

## County Government

The county is governed by three commissioners, who are elected every four years from a slate of four candidates (two Democrats and two Republicans). Other elected officials include County Controller, Sheriff, Coroner, Prothonotary and Clerk of Court, Recorder of Deeds, County Treasurer, and Jury Commissioners.

## Voter Registration

As of November 2008, there are 82,394 registered voters in Lebanon County.[]

- Republican: 44,228 (53.68%)
- Democratic: 27,646 (33.55%)
- Other Parties: 10,520 (12.77%)

## Electoral History

The county has a history of being one of the most Republican counties in Pennsylvania. No Democratic Presidential candidate has won Lebanon County since 1936, when Franklin Delano Roosevelt won re-election in a landslide. Lebanon County was one of only four counties in the state to vote for Barry Goldwater in 1964, along with Snyder, Union, and Wayne counties.

In the 2006 election for U.S. Senate, the county cast 21,756 votes (55.1%) for Republican Rick Santorum and 17,737 (44.9%) for Democrat Bob Casey, Jr., who won the race. In that year's gubernatorial election, the county cast 22,775 votes (57.5%) for Republican Lynn Swann and 16,813 (42.5%) for Democrat Ed Rendell, who won the race.[5]

In the 2004 presidential election, the county cast 37,089 votes (66.6%) for Republican George W. Bush and 18,109 (32.5%) for Democrat John Kerry. In that same year's election for U.S. Senate, the county cast 35,336 votes (66.8%) for Republican Arlen Specter, 13,182 for Democrat Joe Hoeffel, 3,320 (6.3%) for Constitution Party candidate Jim Clymer, and 1,083 (2.0%) for Libertarian Betsy Summers.[6] In the 2008 presidential election the county cast 34,314 votes (58.59%) for Republican John McCain and 23,310 votes (39.8%) for Barack Obama.[7]

In the 2002 gubernatorial election, the county cast 22,659 votes (62.7%) for Republican Mike Fisher and 12,712 (35.2%) for Democrat Ed Rendell, who won the race. In the 2002 race for the U.S. House of Representatives, Republican George Gekas received 21,733 votes (60.9%) from the county while Democrat Tim Holden received 13,945 (39.1%); Holden won.[8]

## Municipalities

Under Pennsylvania law, there are four types of incorporated municipalities: cities, boroughs, townships, and, in at most two cases, towns. The following cities, boroughs and townships are located in Lebanon County:

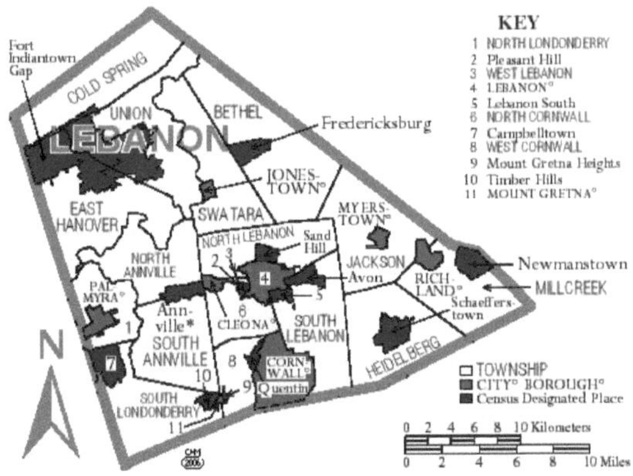


Map of Lebanon County, Pennsylvania with Municipal Labels showing Cities and Boroughs (red), Townships (white), and Census-designated places (blue).

### Cities

- Lebanon is the only incorporated city in Lebanon County.

### Boroughs

- Cleona
- Cornwall
- Jonestown
- Mount Gretna
- Myerstown
- Palmyra
- Richland

### Townships

- Annville Township
- Bethel Township
- Cold Spring Township
- East Hanover Township
- Heidelberg Township
- Jackson Township
- Millcreek Township
- North Annville Township
- North Cornwall Township
- North Lebanon Township
- North Londonderry Township
- South Annville Township
- South Lebanon Township
- South Londonderry Township
- Swatara Township
- Union Township
- West Cornwall Township
- West Lebanon Township

### Census-designated places

Census-designated places are geographical areas designated by the U.S. Census Bureau for the purposes of compiling demographic data. They are not actual jurisdictions under Pennsylvania law. Other unincorporated communities, such as villages, may be listed here as well.

- Annville Township
- Avon
- Campbelltown
- Fort Indiantown Gap
- Fredericksburg
- Hebron
- Lebanon South
- Mount Gretna Heights
- Newmanstown
- Pleasant Hill
- Quentin
- Sand Hill
- Schaefferstown
- Timber Hills

## Education

### Colleges and universities

- Harrisburg Area Community College (Lebanon Campus)
- Lebanon Valley College
- Evangelical Seminary

### Public School Districts

- Annville-Cleona School District
- Cornwall-Lebanon School District
- Eastern Lebanon County School District
- Lebanon School District
- Northern Lebanon School District
- Palmyra Area School District

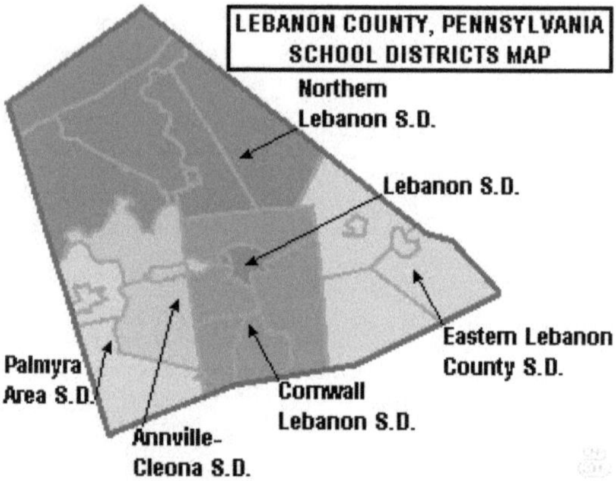

Map of Lebanon County, Pennsylvania School Districts

## See also

- National Register of Historic Places listings in Lebanon County, Pennsylvania

## References

[1] Our County (http://lebanoncountyhistoricalsociety.org/research-archives/lebanon-county). Lebanon County Historical Society. Retrieved on 2013-07-23.

[3] American FactFinder (http://factfinder2.census.gov). Factfinder2.census.gov. Retrieved on 2013-07-23.

[4] University of Virginia Library (http://mapserver.lib.virginia.edu/). Mapserver.lib.virginia.edu. Retrieved on 2013-07-23.

[5] http://www.electionreturns.state.pa.us/ElectionsInformation.aspx?FunctionID=12&ElectionID=24

[6] http://www.electionreturns.state.pa.us/ElectionsInformation.aspx?FunctionID=12&ElectionID=11

[7] http://uselectionatlas.org/RESULTS/statesub.php?year=2008&fips=42075&f=1&off=0&elect=0

[8] http://www.electionreturns.state.pa.us/ElectionsInformation.aspx?FunctionID=12&ElectionID=7

## External links

- County of Lebanon (http://www.lebcounty.org/Pages/default.aspx) (official website)
- Lebanon County Pages (http://www.lebanoncountypages.com/)
- Lebanon County Historical Society (http://www.lebanonhistory.org/)
- Lebanon County Tourism Promotion Agency (http://www.visitlebanoncounty.com/)
- Lebanon Valley Exposition Center and Fairgrounds (http://www.lebexpo.com/)
- (http://www.evangelical.edu)

# United_States

**United States of America**

**Motto:**
"In God we trust" (official)[1] [2] [3]
"E pluribus unum" (Latin) (traditional)
"Out of many, one"

**Anthem:** "The Star-Spangled Banner"

| | | |
|---|---|---|
| **Capital** | | Washington D.C. |
| **Largest city** | | New York City |
| **Official languages** | | None at federal level[a] |
| **National language** | | English[b] |
| **Demonym** | | American |
| **Government** | | Federal presidential constitutional republic |
| - | President | Barack Obama (D) |
| - | Vice President | Joe Biden (D) |

| - | Speaker of the House | John Boehner (R) |
|---|---|---|
| - | Chief Justice | John Roberts |
| **Legislature** | | Congress |
| - | Upper house | Senate |
| - | Lower house | House of Representatives |
| **Independence from the Kingdom of Great Britain** | | |
| - | Declared | July 4, 1776 |
| - | Recognized | September 3, 1783 |
| - | Current constitution | June 21, 1788 |
| **Area** | | |
| - | Total | 9826675 $km^2$[] [c] (3rd/4th)<br>3794101 sq mi |
| - | Water (%) | 6.76 |
| **Population** | | |
| - | 2013 estimate | 316444000[4] (3rd) |
| - | Density | 34.2/$km^2$ (179th)<br>88.6/sq mi |
| **GDP** (PPP) | | 2012 estimate |
| - | Total | $15.685 trillion[] (1st) |
| - | Per capita | $49,922[] (6th) |
| **GDP** (nominal) | | 2012 estimate |
| - | Total | $15.685 trillion[] (1st) |
| - | Per capita | $49,922[] (12th) |
| **Gini** (2011) | | 47.7[5]<br>high · 39th (2009) |
| **HDI** (2013) | | ▲ 0.937[]<br>very high · 3rd |
| **Currency** | | [[]] ($) (USD) |
| **Time zone** | | (UTC−5 to −10) |
| - | Summer (DST) | (UTC−4 to −10[e]) |
| **Drives on the** | | right[g] |
| **Calling code** | | +1 |
| **ISO 3166 code** | | US |
| **Internet TLD** | | .us.gov.mil.edu |
| a. | English is the official language of at least 28 states; some sources give higher figures, based on differing definitions of "official".[] English and Hawaiian are both official languages in the state of Hawaii. | |
| b. | English is the *de facto* language of American government and the sole language spoken at home by 80 percent of Americans aged five and older. Spanish is the second most commonly spoken language. | |

| | |
|---|---|
| c. | Whether the United States or China is larger is disputed. The figure given is from the U.S. Central Intelligence Agency's *The World Factbook*. Other sources give smaller figures. All authoritative calculations of the country's size include only the 50 states and the District of Columbia, not the territories. |
| d. | The population estimate is of people whose usual residence is within the 50 states and the District of Columbia, regardless of nationality. It does not include those living in the territories (over 4 million people, mostly in Puerto Rico). |
| e. | See Time in the United States for details about laws governing time zones in the United States. |
| f. | Does not include insular areas and United States Minor Outlying Islands, which have their own ISO 3166 codes. |
| g. | Except US Virgin Islands. |

The **United States of America** (**USA** or **U.S.A.**), commonly referred to as the **United States** (**US** or **U.S.**) and **America**, is a federal republic[6] [7] consisting of 50 states and a federal district. The 48 contiguous states and the federal district of Washington, D.C. are in central North America between Canada and Mexico. The state of Alaska is the northwestern part of North America, west of Canada and east of Russia which is across the Bering Strait in Asia, and the state of Hawaii is an archipelago in the mid-North Pacific. The country also has five populated and nine unpopulated territories in the Pacific and the Caribbean.

At 3.79 million square miles (9.83 million km$^2$) in total and with around 315 million people, the United States is the third largest country by land area and population. It is one of the world's most ethnically diverse and multicultural nations, the product of large-scale immigration from many countries.[8] The geography and climate of the U.S. is also extremely diverse, with deserts, plains, forests, and mountains that are home to a wide variety of wildlife.

Paleo-indians migrated from Asia to what is now the United States mainland around 12,000 years ago. European colonization began around 1600 and came mostly from England. The United States emerged from 13 British colonies located along the Atlantic seaboard. Disputes between Great Britain and the American colonies led to the American Revolution. On July 4, 1776, delegates from the 13 colonies unanimously issued the Declaration of Independence, which established the United States of America. The American Revolutionary War, which ended with the recognition of independence of the United States from the Kingdom of Great Britain, was the first successful war of independence against a European colonial empire.[9] [10] The current Constitution was adopted on September 17, 1787; 27 Amendments have since been added to the Constitution. The first 10 amendments, collectively named the Bill of Rights, were ratified in 1791 and guarantee many fundamental civil rights and freedoms.

Driven by the doctrine of manifest destiny, the United States embarked on a vigorous expansion across North America throughout the 19th century.[] This involved displacing native tribes, acquiring new territories, and gradually admitting new states.[] The American Civil War ended legalized slavery in the United States.[11] By the end of the 19th century, the United States extended into the Pacific Ocean,[] and its economy was the world's largest.[12] The Spanish–American War and World War I confirmed the country's status as a global military power. The United States emerged from World War II as a global superpower, the first country with nuclear weapons, and a permanent member of the United Nations Security Council. The end of the Cold War and the dissolution of the Soviet Union left the United States as the sole superpower.

The United States is a developed country and has the world's largest national economy, with an estimated 2013 GDP of $16.2 trillion –22% of global GDP at purchasing-power parity, as of 2011.[] [] [13] The per capita GDP of the U.S. was the world's sixth-highest as of 2010.[] A 2013 study ranked the US as having the fourth most unequal income distribution among OECD nations,[14] though it has the highest mean and second highest median household income in the OECD, as well as the highest average employee compensation.[] [] The economy is fueled by an abundance of natural resources, a well-developed infrastructure,[15] and high productivity;[16] and while its economy is considered post-industrial it continues to be one of the world's largest manufacturers.[17] The country accounts for 39% of global military spending,[18] being the foremost economic and military power, a prominent political and cultural force in the world, and a leader in scientific research and technological innovation.[19] [20]

## Etymology

In 1507, German cartographer Martin Waldseemüller produced a world map on which he named the lands of the Western Hemisphere "America" after Italian explorer and cartographer Amerigo Vespucci.[21]

The first documentary evidence of the phrase "United States of America" is from a letter dated January 2, 1776, written by Stephen Moylan, Esq., George Washingon's aide-de-camp and Muster-Master General of the Continental Army. Addressed to Col. Joseph Reed, Moylan expressed his wish to carry the "full and ample powers of the United States of America" to Spain to assist in the revolutionary war effort.[22]

The first publicly published evidence of the phrase "United States of America" was in an anonymously written essay in *The Virginia Gazette* newspaper in Williamsburg, Virginia on April 6, 1776.[23] [24] In June 1776, Thomas Jefferson included the phrase "UNITED STATES OF AMERICA" in all capitalized letters in the headline of his "original Rough draught" of the Declaration of Independence.[25] [26] In the final Fourth of July version of the Declaration, the pertinent section of the title was changed to read, "The unanimous Declaration of the thirteen united States of America".[27]

In 1777 the Articles of Confederation announced, "The Stile of this Confederacy shall be 'The United States of America'".[28]

The short form "United States" is also standard. Other common forms include the "U.S.", the "USA", and "America". Colloquial names include the "U.S. of A." and, internationally, the "States". "Columbia", a name popular in poetry and songs of the late 1700s,[29] derives its origin from Christopher Columbus; it appears in the name "District of Columbia".

The standard way to refer to a citizen of the United States is as an "American". "United States", "American" and "U.S." are used to refer to the country adjectivally ("American values", "U.S. forces"). "American" is rarely used in English to refer to subjects not connected with the United States.[30]

The phrase "United States" was originally treated as plural, a description of a collection of independent states—e.g., "the United States are"—including in the Thirteenth Amendment to the United States Constitution, ratified in 1865. It became common to treat it as singular, a single unit—e.g., "the United States is"—after the end of the Civil War. The singular form is now standard; the plural form is retained in the idiom "these United States".[] The difference has been described as more significant than one of usage, but reflecting the difference between a collection of states and a unit.[31]

In non-English languages, the name is frequently translated as the translation of either the "United States" or "United States of America", and colloquially as "America". In addition, an initialism is sometimes used.[32]

# History

## Native American and European contact

1766 engraving of a meeting between Native Americans and Europeans

People from Asia migrated to the North American continent approximately 12,000 or more years ago.[33] Some, such as the pre-Columbian Mississippian culture, developed advanced agriculture, grand architecture, and state-level societies. After European explorers and traders made the first contacts, it is estimated that their population declined due to various reasons, including diseases such as smallpox and measles to which indigenous Americans had no natural immunities,[34] [35] intermarriage,[36] and violence.[37] [38] [39]

In the early days of colonization many settlers were subject to shortages of food, disease and attacks from native Indians. Indians were also often at war with neighboring tribes and would often enslave their defeated enemy, a practice that was also soon used by various colonists who captured Indians in battle. During the various colonial wars, many colonists were also captured by Indians as slaves and taken north to Canada and sold to the French.[40]

At the same time however many natives and settlers got along and came to depend on each other, especially settlers during the winter months. Natives also came to depend on settlers for guns, ammunition, powder and other modern devices. Because many tribes were frequently at war with one another it became imperative to establish and secure good relationships with at least one group of colonists. As colonists began to spread out into the interior their contact with native Indians increased, sometimes resulting in good relations, oftentimes resulting in conflict. In the process "Native American influenced colonist, and colonist influenced Native American".[41]

Natives taught many settlers where, when and how to hunt and fish in the vast frontier that lay before them whose elements were generally unknown to the Europeans. In order to survive settlers often depended on native Indians who taught them how to adopt to the Indian's "hunting culture" and learned the use of animal skins as camouflage, decoys along with various whistles and calls used to attract prey. European ministries and others felt it was important to "civilize" the Indians and urged them to concentrate on farming and ranching and not depend primarily on hunting and gathering. At the same time Indians offered the benefit of their experience in growing corn, an unknown crop in Europe, and in the use of dead fish and other methods as fertilizer. It was not long before many Indians began to grow new crops and raise livestock and poultry in their communities and made use of the various living utilities settlers had to offer.[42] [43]

Initially the Puritan and Wampanoag were peaceful, however the King Philip's War began following cultural and religious differences between the colonists and the Wampanoag;[44] [45] by the war's end, the European colonists had defeated the Native Americans and were able to expand and control New England.[46] In Carolina, Native Americans were captured and sold into slavery to both New England and the West Indies. In 1676, the Virginia colony legally sanctioned the enslavement of Native Americans.[47] Conversely, the Five Civilized Tribes were involved in the institution of African slavery as planters.[48]

## Settlements

Further information: European colonization of the Americas and 13 colonies

The signing of the Mayflower Compact, 1620.

After Columbus' discovery of the New World in 1492 other explorers followed.[49] The first Spanish explorers landed in "La Florida" in 1513. Conquistadors explored much of the continent's interior and Spain later set up some settlements in parts of Florida and the American southwest that were eventually merged into the United States.[50] There were also some French attempts to colonize the east coast, and later more successful settlements along the Mississippi River. Many early European colonies failed due to starvation, disease, harsh weather, Indian attacks, or warfare with European rivals. The fate of the "lost" English colony of Roanoke in the 1580s is an enduring mystery.

James I on April 10, 1606 chartered The Virginia Company with the purpose of establishing English settlements on the eastern coast of North America. The Virginia Colony was planted in 1607 with Jamestown and the Pilgrims' Plymouth Colony in 1620. Both colonies suffered initial hardships and great loss of life, but eventually stabilized and became the first successful English settlements in America. Both also saw efficiency greatly improve when personal property replaced the early communal operations.[51] The continent's first elected legislative assembly, Virginia's House of Burgesses created in 1619, and the Mayflower Compact, signed by the Pilgrims before disembarking, established precedents for the pattern of representative self-government and constitutionalism that would develop throughout the American colonies.[52] [53] Tens of thousands of Puritans later settled New England.

Other New England colonies were established. Much of the territory between them and Virginia was controlled by the Dutch until England seized it in the late 17th century during the Anglo-Dutch Wars, leading to the creation of the Middle Colonies.[54] Trade with and Christian evangelism to local Indians were established in the colonies' early days, though relations would alternate from friendly to tense, and were characterized by periodic bouts of warfare, often with some tribes allying themselves with the English against common foes. Incidents like the massacre of 1622, the Pequot War, and King Philip's War caused great destruction and threatened the existence of entire colonies, but resulted in reprisals that ultimately saw the power of enemy tribes reduced or broken, facilitating the expansion of English settlements.[55] [56]

Most settlers in every colony were small farmers, but other industries developed. Tobacco was popular in Europe and became a major early cash crop. Furs, fishing, lumber, rum, rice, indigo, construction, wheat, ranching, and eventually shipbuilding contributed to economic growth. By the late colonial period Americans were producing one-seventh of the world's iron supply.[57] Cities eventually dotted the coast to support local economies and serve as trade hubs. English colonists were supplemented by waves of Scotch-Irish and other groups. As coastal land grew more expensive people pushed west into the hills and backwoods, seeking to carve an existence out of virgin wilderness.[58]

Settlers were a diverse mix of adventurers, profit seekers, people wanting religious freedom, and those who simply saw an opportunity for a better life.[59] Many came as indentured servants, either convicts or people who otherwise couldn't afford passage voluntarily signing contracts, and were set free after completing their specified term of service. Two-thirds of all Virginia settlers between 1630 and 1680 arrived indentured.[60]

Penn's Treaty between the Quaker settlers and Native Americans was never violated during William Penn's lifetime.

The first African slaves were brought to the Americas by Spanish conquistadors in the 1500s shortly after Columbus' voyages. Most slaves were shipped to sugar colonies in the Caribbean and to Brazil, where life expectancy was about seven years.[61] Life expectancy was much higher in North America because of less disease and better food and treatment, so the numbers of slaves grew rapidly into the millions by excesses of births over deaths.[62] [63]

Colonial society was largely divided over the religious and moral implications of slavery and many colonies passed acts for and against the practice.[64] [65] By the turn of the 18th century, African slaves were becoming the primary source of bonded labor in many regions.[66] Some colonists participated in the lucrative, slave oriented "Golden Triangle", involving planters, merchants of various types, shippers, and the African tribal chiefs who provided them with slaves.[47] [67]

With the 1729 division of the Carolinas and the 1732 colonization of Georgia, the 13 colonies that would become the United States of America were established.[] All had local governments with elections open to most free men, with a growing devotion to the ancient rights of Englishmen and a sense of self-government stimulating support for republicanism.[] With extremely high birth rates, low death rates, and steady settlement, the colonial population grew rapidly. Relatively small Indian populations were eclipsed.[68] The Christian revivalist movement of the 1730s and 1740s known as the Great Awakening fueled interest in both religion and religious liberty.

In the French and Indian War, British forces seized Canada from the French, but the francophone population remained politically isolated from the southern colonies. Excluding the Native Americans, who were being conquered and displaced, those 13 colonies had a population of over 2.1 million in 1770, about one-third that of Britain. Despite continuing new arrivals, the rate of natural increase was such that by the 1770s only a small minority of Americans had been born overseas.[69] The colonies' distance from Britain had allowed the development of self-government, but their success motivated monarchs to periodically seek to reassert Royal authority.

## Independence and expansion

Further information: American Revolutionary War, Declaration of Independence, and American Revolution

The American Revolution was the first successful colonial war of independence against a European power. Americans had developed a democratic system of local government and an ideology of "republicanism" that held government rested on the will of the people (not the king), which strongly opposed corruption and demanded civic virtue. They demanded their rights as Englishmen and rejected British efforts to impose taxes without the approval of colonial legislatures. The British insisted and the conflict escalated to full-scale war in 1775, the American Revolutionary War.[] On June 14, 1775, the Continental Congress, convening in Philadelphia, established a Continental Army under the command of George Washington.[] Proclaiming that "all men are created equal" and endowed with "certain unalienable Rights", the Congress adopted the Declaration of Independence, drafted largely by Thomas Jefferson, on July 4, 1776. That date is now celebrated annually as America's Independence Day. In

1777, the Articles of Confederation established a weak government that operated until 1789.[]

The *Declaration of Independence*: the Committee of Five presenting their draft to the Second Continental Congress in 1776.

U.S. Territories were acquired, organized with citizens, republican self-government, then admitted states on equal footing.

After a naval victory followed by the British defeat at Yorktown by American forces assisted by the French,[70] the United States was independent. In the peace treaty of 1783 Britain recognized American sovereignty over most territory east of the Mississippi River. Nationalists calling for a much stronger federal government with powers of taxation led the constitutional convention in 1787. After intense debate in state conventions the United States Constitution was ratified in 1788. The first Senate, House of Representatives, and president—George Washington—took office in 1789. The Bill of Rights, forbidding federal restriction of personal freedoms and guaranteeing a range of legal protections, was adopted in 1791.[71]

Attitudes toward slavery were shifting; nearly all states officially outlawed the international slave trade before the federal government criminalized it in 1808.[] Slavery had become more pronounced in the south than the north because the land there was better suited for large scale cash crop cultivation than the rocky ground and cooler climate of New England.[72] [73] All the Northern states abolished slavery between 1780 and 1804, leaving the slave states of the South as defenders of the "peculiar institution". With cotton a highly profitable plantation crop after 1820, slave interests in the Southern states maintained that slavery was a positive good for everyone, including the slaves.[] The Second Great Awakening, beginning about 1800, converted millions to evangelical Protestantism. In the North it energized multiple social reform movements, including abolitionism.[]

Americans' eagerness to expand westward prompted a long series of Indian Wars.[] The Louisiana Purchase of French-claimed territory under President Thomas Jefferson in 1803 almost doubled the nation's size.[74] The War of 1812, declared against Britain over various grievances and fought to a draw, strengthened U.S. nationalism.[] A series of U.S. military incursions into Florida led Spain to cede it and other Gulf Coast territory in 1819.[]

President Andrew Jackson took office in 1829, and began a set of reforms which led to the era of Jacksonian democracy, which is considered to have lasted from 1830 to 1850. This included many reforms, such as wider male suffrage, and various adjustments to the power of the Federal government. This also led to the rise of the Second Party System, which refers to the dominant parties which existed from 1828 to 1854.

The Trail of Tears in the 1830s exemplified the Indian removal policy that moved Indians to their own reservations, sometimes by force, with small annual government subsidies. The United States annexed the Republic of Texas in 1845, amid a period when the concept of Manifest Destiny was becoming popular.[] The 1846 Oregon Treaty with Britain led to U.S. control of the present-day American Northwest.[] The U.S. victory in the Mexican-American War resulted in the 1848 cession of California and much of the present-day American Southwest.[]

The California Gold Rush of 1848–49 further spurred western migration.[] New railways made relocation easier for settlers and increased conflicts with Native Americans.[] Over a half-century, up to 40 million American bison, or buffalo, were slaughtered for skins and meat and to ease the railways' spread.[] The loss of the buffalo, a primary resource for the plains Indians, was an existential blow to many native cultures.[] In 1869, President Ulysses S. Grant's Peace policy reversed the previous costly policy of "wars of extermination" in order to civilize and give Indians eventual United State citizenship having incorporated Indians as wards of the state, led by a philanthropic Board of Indian Commissioners.[75]

## Civil War and Reconstruction Era

Further information: American Civil War and Reconstruction Era

Starting in the 1780s inherent divisions between the North and the South in American society over slavery ultimately led to the American Civil War.[76] Initially, the Founders of the nation had been able to keep the Union solvent by compromises worked out at the Constitutional Convention and to remain a single nation.[76]

Battle of Gettysburg, Pennsylvania. The American Civil WarCivil War cemented the Union and spurred the steel industry and First Transcontinental Railroadintercontinental railroad construction.

During the years leading up to the American Civil War tensions between slave and free states mounted with arguments about the relationship between the state and federal governments, as well as violent conflicts over the spread of slavery into new states.[] Abraham Lincoln, candidate of the largely antislavery Republican Party, was elected president in 1860.[] Before he took office, seven slave states declared their secession—which the federal government maintained was illegal—and formed the Confederate States of America.[]

With the Confederate attack upon Fort Sumter, the Civil War began and four more slave states joined the Confederacy.[] Lincoln's Emancipation Proclamation in 1863 declared slaves in the Confederacy to be free, though not those in Union slave states. Following the Union victory in 1865, three amendments to the U.S. Constitution ensured freedom for the nearly four million African Americans who had been slaves,[77] made them citizens, and gave them voting rights. The war and its resolution led to a substantial increase in federal power.[78] The war remains the deadliest conflict in American history, resulting in the deaths of 620,000 soldiers.[79]

The assassination of Abraham Lincoln radicalized Republican Reconstruction policies aimed at reintegrating and rebuilding the Southern states while ensuring the rights of the newly freed slaves.[] President Ulysses S. Grant implemented the Department of Justice and used the U.S. Military to enforce suffrage and civil rights for African Americans in the South destroying the Ku Klux Klan in 1871 under the Force Acts.[80] The resolution of the disputed 1876 presidential election by the Compromise of 1877 ended Reconstruction; Jim Crow laws soon disenfranchised many African Americans.[]

## Industrialization

In the North, urbanization and an unprecedented influx of immigrants from Southern and Eastern Europe hastened the country's industrialization. The wave of immigration, lasting until 1924, provided labor and transformed American culture.[] United States immigration policies were Eurocentric, which barred Asians from naturalization, and restricted their immigration beginning with the Chinese Exclusion Act in 1882.[81] National infrastructure development spurred economic growth. The end of the Civil War spurred greater settlement and development of the American Old West. This was due to a variety of social and technological developments, including the completion of the First Transcontinental Telegraph in 1861 and the First Transcontinental Railroad in 1869.

Ellis Island, New York City. East Coast of the United StatesEast Coast immigrants worked in factories, railroads, and mines, and created demand for industrialized agriculture.

The 1867 Alaska Purchase from Russia completed the country's mainland expansion. The Wounded Knee Massacre in 1890 was the last major armed conflict of the Indian Wars. In 1893, the indigenous monarchy of the Pacific Kingdom of Hawaii was overthrown in a coup led by American residents; the United States annexed the archipelago in 1898. Victory in the Spanish–American War the same year demonstrated that the United States was a world power and led to the annexation of Puerto Rico, Guam, and the Philippines.[82] The Philippines gained independence a half-century later; Puerto Rico and Guam remain U.S. territories.

The emergence of many prominent industrialists at the end of the 19th century gave rise to the Gilded Age, a period of growing affluence and power among the business class. The hardships imposed upon the working classes during this period led to the rise of anarchist and socialist movements in the US.[83] In 1914 alone, 35,000 workers died in industrial accidents and 700,000 were injured.[84] This period eventually ended with the beginning of the Progressive Era, a period of great reforms in many societal areas, including regulatory protection for the public, greater antitrust measures, and attention to living conditions for the working classes. President Theodore Roosevelt was one leading proponent of progressive reforms.

## World War I, Great Depression, and World War II

Further information: World War I, Great Depression, and World War II

At the outbreak of World War I in 1914, the United States remained neutral. Most Americans sympathized with the British and French, although many opposed intervention.[85] In 1917, the United States joined the Allies, and the American Expeditionary Forces helped to turn the tide against the Central Powers. President Woodrow Wilson took a leading diplomatic role at the Paris Peace Conference of 1919 which helped to shape the post-war world. Wilson advocated strongly for the U.S. to join the League of Nations. However, the Senate refused to approve this, and did not ratify the Treaty of Versailles, which established the League of Nations.[86]

The Dust Bowl brought agricultural depression, impacted industrial markets, and led to large relocation out of the Great Plains.

WWII invasion of Europe required war industry, accelerating migration to big cities and large scale manufacturing

The country pursued a policy of unilateralism, verging on isolationism.[86] In 1920, the women's rights movement, led by Carrie Chapman Catt, won passage of a constitutional amendment granting women's suffrage.[] The prosperity of the Roaring Twenties ended with the Wall Street Crash of 1929 that triggered the Great Depression.

After his election as president in 1932, Franklin D. Roosevelt responded with the New Deal, a range of policies increasing government intervention in the economy, including the establishment of the Social Security system.[87] The Dust Bowl of the mid-1930s impoverished many farming communities and spurred a new wave of western migration.

The United States, effectively neutral during World War II's early stages after Nazi Germany's invasion of Poland in September 1939, began supplying material to the Allies in March 1941 through the Lend-Lease program. On December 7, 1941, the Empire of Japan launched a surprise attack on Pearl Harbor, prompting the United States to join the Allies against the Axis powers as well as the internment of Japanese Americans by the thousands.[88] Participation in the war spurred capital investment and industrial capacity, and the production figures after the Americans started to unfold the awesome productive capacity of their economy became the stuff of legend.[89] Though the nation lost more than 400,000 soldiers,[90] among the major combatants, the United States was the only nation to become richer because of the war.[91]

Allied conferences at Bretton Woods and Yalta outlined a new system of international organizations that placed the United States and Soviet Union at the center of world affairs. As victory was won in Europe, a 1945 international conference held in San Francisco produced the United Nations Charter, which became active after the war.[92] The United States, having developed the first nuclear weapons, used them on the Japanese cities of Hiroshima and Nagasaki in August. Japan surrendered on September 2, ending the war.[93]

## Cold War and Civil Rights era

Civil Rights leaders, including Ralph Abernathy and Martin Luther King, Jr., lead one of the Selma to Montgomery marches

A vehicle enters West Berlin through the Western Allies' Checkpoint Charlie after the fall of the Berlin Wall, marking the beginning of the end of the Cold War

The United States and the Soviet Union jockeyed for power after World War II during the Cold War, dominating the military affairs of Europe through NATO and the Warsaw Pact, respectively. While they engaged in proxy wars and developed powerful nuclear arsenals, the two countries avoided direct military conflict. The U.S. often opposed

Third World left-wing movements that it viewed as Soviet-sponsored. American troops fought Communist Chinese and North Korean forces in the Korean War of 1950–53. The House Un-American Activities Committee pursued a series of investigations into suspected leftist subversion, while Senator Joseph McCarthy became the figurehead of anticommunist sentiment.[94]

The 1961 Soviet launch of the first manned spaceflight prompted President John F. Kennedy's call for the United States to be first to land "a man on the moon", achieved in 1969.[95] Kennedy also faced a tense nuclear showdown with Soviet forces in Cuba.[96] Meanwhile, the United States experienced sustained economic expansion. Amidst the presence of various white nationalist groups, particularly the Ku Klux Klan, a growing civil rights movement used nonviolence to confront segregation and discrimination. This was symbolized and led by black Americans such as Rosa Parks and Martin Luther King, Jr. On the other hand, some black nationalist groups such as the Black Panther Party and Malcolm X had a more militant scope.

Following Kennedy's assassination in 1963, the Civil Rights Act of 1964, Voting Rights Act of 1965, and Immigration and Nationality Act of 1965 were passed under President Lyndon B. Johnson.[97] He also signed into law the Medicare and Medicaid programs.[98] Johnson also expanded a proxy war in Southeast Asia into the ultimately unsuccessful Vietnam War. A widespread countercultural movement grew, fueled by opposition to the war, black nationalism, and the sexual revolution. Betty Friedan, Gloria Steinem, and others led a new wave of feminism that sought political, social, and economic equality for women.

As president, Richard Nixon ended U.S. involvement in the Vietnam War, improved relations with China, and oversaw the beginning of a period of generally eased relations with the Soviet Union. As a result of the Watergate scandal, in 1974 Nixon became the first U.S. president to resign, to avoid being impeached on charges including obstruction of justice and abuse of power. The Jimmy Carter administration of the late 1970s was marked by stagflation and the Iran hostage crisis. The election of Ronald Reagan as president in 1980 heralded a rightward shift in American politics,[99] [100] [101] [102] reflected in major changes in taxation and spending priorities.[103] His second term in office brought both the Iran–Contra scandal and significant diplomatic progress with the Soviet Union.[104] The subsequent Soviet collapse ended the Cold War.[105] [106] [107] [108] [109]

## Contemporary era

Under President George H. W. Bush, the United States took a lead role in the UN–sanctioned Gulf War.[110] The longest economic expansion in modern U.S. history—from March 1991 to March 2001—encompassed the Bill Clinton administration and the dot-com bubble.[111]

September 11, 2001 attacks on the World Trade Center in New York City.

On September 11, 2001, al-Qaeda terrorists under the leadership of Osama bin Laden struck the World Trade Center in New York City and The Pentagon near Washington, D.C., killing nearly 3,000 people.[112] In response, the George W. Bush administration launched the global War on Terror, invading Afghanistan and removing the Taliban government and al-Qaeda training camps.[113] Bin Laden was killed by a Navy SEAL raid in Pakistan in 2011;[114] Taliban insurgents continue to fight a guerrilla war.[115] In 2003, the United States and several allied forces invaded Iraq to engineer regime change there; the war lasted eight years.[116] [117] [118] In 2005, Hurricane Katrina caused severe destruction along much of the Gulf Coast, devastating New Orleans.[119]

In 2008, amid a global economic recession, the first African-American president, Barack Obama, was elected.[120]

## Geography, climate, and environment

The land area of the contiguous United States is 2959064 square miles ( $km^2$). Alaska, separated from the contiguous United States by Canada, is the largest state at 663268 square miles ( $km^2$). Hawaii, occupying an archipelago in the central Pacific, southwest of North America, is 10931 square miles ( $km^2$) in area.[121]

A composite satellite image of the contiguous United States

The United States is the world's third or fourth largest nation by total area (land and water), ranking behind Russia and Canada and just above or below China. The ranking varies depending on how two territories disputed by China and India are counted and how the total size of the United States is measured: calculations range from 3676486 square miles ( $km^2$)[122] to 3717813 square miles ( $km^2$)[123] to 3794101 square miles ( $km^2$).[] Measured by only land area, the United States is third in size behind Russia and China, just ahead of Canada.[124]

The coastal plain of the Atlantic seaboard gives way further inland to deciduous forests and the rolling hills of the Piedmont. The Appalachian Mountains divide the eastern seaboard from the Great Lakes and the grasslands of the Midwest. The Mississippi–Missouri River, the world's fourth longest river system, runs mainly north–south through the heart of the country. The flat, fertile prairie of the Great Plains stretches to the west, interrupted by a highland region in the southeast.

The Rocky Mountains, at the western edge of the Great Plains, extend north to south across the country, reaching altitudes higher than 14,000 feet (4,300 m) in Colorado. Farther west are the rocky Great Basin and deserts such as the Chihuahua and Mojave. The Sierra Nevada and Cascade mountain ranges run close to the Pacific coast, both ranges reaching altitudes higher than 14000 feet ( m).

A Bald Eagle, the national bird of the United States since 1782.

The lowest and highest points in the continental United States are in the state of California, and only about 80 miles ( km) apart. At 20,320 feet (6,194 m), Alaska's Mount McKinley is the tallest peak in the country and in North America. Active volcanoes are common throughout Alaska's Alexander and Aleutian Islands, and Hawaii consists of volcanic islands. The supervolcano underlying Yellowstone National Park in the Rockies is the continent's largest volcanic feature.[125]

The United States, with its large size and geographic variety, includes most climate types. To the east of the 100th meridian, the climate ranges from humid continental in the north to humid subtropical in the south. The southern tip of Florida is tropical, as is Hawaii. The Great Plains west of the 100th meridian are semi-arid. Much of the Western mountains are alpine. The climate is arid in the Great Basin, desert in the Southwest, Mediterranean in coastal California, and oceanic in coastal Oregon and Washington and southern Alaska. Most of Alaska is subarctic or polar. Extreme weather is not uncommon—the states bordering the Gulf of Mexico are prone to hurricanes, and most of the world's tornadoes occur within the country, mainly in the Midwest's Tornado Alley.[126]

The U.S. ecology is considered "megadiverse": about 17,000 species of vascular plants occur in the contiguous United States and Alaska, and over 1,800 species of flowering plants are found in Hawaii, few of which occur on the mainland.[127] The United States is home to more than 400 mammal, 750 bird, and 500 reptile and amphibian species.[128] About 91,000 insect species have been described.[129]

There are 58 national parks and hundreds of other federally managed parks, forests, and wilderness areas.[130] Altogether, the government owns 28.8% of the country's land area.[] Most of this is protected, though some is leased for oil and gas drilling, mining, logging, or cattle ranching; 2.4% is used for military purposes.[] [131] []

### Environmental issues

Environmental issues have been high on the national agenda since 1970. The controversies include debates on oil and nuclear energy, dealing with air and water pollution, the economic costs of protecting wildlife, and international responses to global warming.[132] [133]

Many federal and state agencies are involved. The most prominent is the Environmental Protection Agency (EPA), created by presidential order in 1970.[134] The idea of wilderness has shaped the management of public lands since 1964, with the Wilderness Act.[135] The Endangered Species Act of 1973 protects threatened and endangered species and their habitats, which are monitored by the United States Fish and Wildlife Service.

## Demographics

### Population

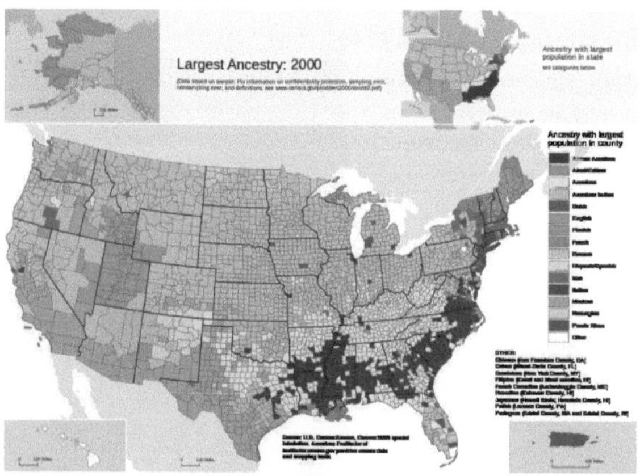


Largest ancestry groups by county, 2000

| Race/Ethnicity | |
|---|---|
| (as given by the 2010 Census)[] | |
| **By race:** | |
| White | 72.4% |
| Black American | 12.6% |
| Asian | 4.8% |
| American Indian and Alaska Native | 0.9% |
| Native Hawaiian and Pacific Islander | 0.2% |
| Other | 6.2% |
| Multiracial (2 or more) | 2.9% |
| **By ethnicity:[136]** | |
| Hispanic/Latino (of any race) | 16.3% |
| Non-Hispanic/Latino (of any race) | 83.7% |

The U.S. Census Bureau estimates the country's population now to be 316444000,[4] including an approximate 11.2 million illegal aliens.[137] The U.S. population almost quadrupled during the 20th century, from about 76 million in 1900.[138] The third most populous nation in the world, after China and India, the United States is the only major industrialized nation in which large population increases are projected.[]

With a birth rate of 13 per 1,000, 35% below the world average, its population growth rate is positive at 0.9%, significantly higher than those of many developed nations.[139] In fiscal year 2012, over one million immigrants (most of whom entered through family reunification) were granted legal residence.[140] Mexico has been the leading source of new residents for over two decades; since 1998, China, India, and the Philippines have been in the top four sending countries every year.[141] [142] 9 million Americans identify as lesbian, gay, bisexual or transgender, making up four percent of the population.[143] A 2010 survey found that seven percent of men and eight percent of women identified as gay, lesbian or bisexual.[144]

The United States has a very diverse population—31 ancestry groups have more than one million members.[] White Americans are the largest racial group; German Americans, Irish Americans, and English Americans constitute three of the country's four largest ancestry groups.[] Black Americans are the nation's largest racial minority and third largest ancestry group.[] Asian Americans are the country's second largest racial minority; the three largest Asian American ethnic groups are Chinese Americans, Filipino Americans, and Indian Americans.[]

In 2010, the U.S. population included an estimated 5.2 million people with some American Indian or Alaska Native ancestry (2.9 million exclusively of such ancestry) and 1.2 million with some native Hawaiian or Pacific island ancestry (0.5 million exclusively).[] The census counted more than 19 million people of "Some Other Race" who were "unable to identify with any" of its five official race categories in 2010.[]

The population growth of Hispanic and Latino Americans (the terms are officially interchangeable) is a major demographic trend. The 50.5 million Americans of Hispanic descent[] are identified as sharing a distinct "ethnicity" by the Census Bureau; 64% of Hispanic Americans are of Mexican descent.[] Between 2000 and 2010, the country's Hispanic population increased 43% while the non-Hispanic population rose just 4.9%.[] Much of this growth is from immigration; as of 2007, 12.6% of the U.S. population was foreign-born, with 54% of that figure born in Latin America.[145]

Fertility is also a factor; as of 2010 the average Hispanic (of any race) woman gave birth to 2.35 children in her lifetime, compared to 1.97 for non-Hispanic black women and 1.79 for non-Hispanic white women (both below the replacement rate of 2.1).[] Minorities (as defined by the Census Bureau as all those beside non-Hispanic, non-multiracial whites) constituted 36.3% of the population in 2010,[146] and over 50% of children under age one,[147] and are projected to constitute the majority by 2042.[148] This contradicts the report by the National Vital Statistics Reports, based on the U.S. census data, which concludes that, 54% (2,162,406 out of 3,999,386 in 2010) of births were non-Hispanic white.[]

About 82% of Americans live in urban areas (including suburbs);[] about half of those reside in cities with populations over 50,000.[149] In 2008, 273 incorporated places had populations over 100,000, nine cities had more than one million residents, and four global cities had over two million (New York City, Los Angeles, Chicago, and Houston).[]

There are 52 metropolitan areas with populations greater than one million.[] Of the 50 fastest-growing metro areas, 47 are in the West or South.[150] The metro areas of Dallas, Houston, Atlanta, and Phoenix all grew by more than a million people between 2000 and 2008.[]

**Leading population centers (see complete list)**

| Rank | Core city (cities) | Metro area population | Metropolitan Statistical Area | Region[151] |
|---|---|---|---|---|
| 1 | **New York City** | 19,015,900 | New York–New Jersey–Connecticut–Pennsylvania, NY–NJ–CT–PA MSA | Northeast |
| 2 | **Los Angeles** | 12,944,801 | Los Angeles–Long Beach–Santa Ana, CA MSA | West |
| 3 | **Chicago** | 9,504,753 | Chicago–Joliet–Naperville, IL–IN–WI MSA | Midwest |
| 4 | **Dallas–Fort Worth** | 6,526,548 | Dallas–Fort Worth–Arlington, TX MSA | South |
| 5 | **Houston** | 6,086,538 | Houston–The Woodlands-Sugar Land MSA | South |
| 6 | **Philadelphia** | 5,992,414 | Philadelphia–Camden–Wilmington, PA–NJ–DE–MD MSA | Northeast |
| 7 | **Washington, D.C.** | 5,703,948 | Washington, DC–VA–MD–WV MSA | Northeast |
| 8 | **Miami** | 5,670,125 | Miami–Fort Lauderdale–Pompano Beach, FL MSA | South |
| 9 | **Atlanta** | 5,359,205 | Atlanta–Sandy Springs–Marietta, GA MSA | South |
| 10 | **Boston** | 4,591,112 | Boston–Cambridge–Quincy, MA–NH MSA | Northeast |
| 11 | **San Francisco** | 4,391,037 | San Francisco–Oakland–Fremont, CA MSA | West |
| 12 | **San Bernardino-Riverside** | 4,304,997 | San Bernandino–Riverside–Ontario, CA MSA | West |
| 13 | **Detroit** | 4,285,832 | Detroit–Warren–Livonia, MI MSA | Midwest |
| 14 | **Phoenix** | 4,263,236 | Phoenix–Mesa–Glendale, AZ MSA | West |
| 15 | **Seattle** | 3,500,026 | Seattle–Tacoma–Bellevue, WA MSA | West |
| 16 | **Minneapolis–St. Paul** | 3,318,486 | Minneapolis–St. Paul–Bloomington, MN–WI MSA | Midwest |
| 17 | **San Diego** | 3,140,069 | San Diego–Carlsbad–San Marcos, CA MSA | West |
| 18 | **Tampa–St. Petersburg** | 2,824,724 | Tampa–St. Petersburg–Clearwater, FL MSA | South |
| 19 | **St. Louis** | 2,817,355 | St. Louis–St. Charles–Farmington, MO–IL MSA | Midwest |
| 20 | **Baltimore** | 2,729,110 | Baltimore–Towson, MD MSA | Northeast |

based upon 2011 population estimates from the U.S. Census Bureau[152]

New York City

Los Angeles

Chicago

## Language

| Languages (2010)[153] | |
|---|---|
| English (*only*) | 229.7 million |
| Spanish, incl. Creole | 37.0 million |
| Chinese | 2.8 million |
| French, incl. Creole | 2.1 million |
| Tagalog | 1.6 million |
| Vietnamese | 1.4 million |
| Italian | 1.1 million |
| Korean | 1.1 million |
| German | 1.1 million |

English (American English) is the de facto national language. Although there is no official language at the federal level, some laws—such as U.S. naturalization requirements—standardize English. In 2010, about 230 million, or 80% of the population aged five years and older, spoke only English at home. Spanish, spoken by 12% of the population at home, is the second most common language and the most widely taught second language.[153] [154] Some Americans advocate making English the country's official language, as it is in at least 28 states.[]

Both Hawaiian and English are official languages in Hawaii, by state law.[155] While neither has an official language, New Mexico has laws providing for the use of both English and Spanish, as Louisiana does for English and French.[156] Other states, such as California, mandate the publication of Spanish versions of certain government documents including court forms.[157] Many jurisdictions with large numbers of non-English speakers produce government materials, especially voting information, in the most commonly spoken languages in those jurisdictions.

Several insular territories grant official recognition to their native languages, along with English: Samoan and Chamorro are recognized by American Samoa and Guam, respectively; Carolinian and Chamorro are recognized by the Northern Mariana Islands; Spanish is an official language of Puerto Rico and is more widely spoken than English there.

## Religion

The First Amendment of the U.S. Constitution guarantees the free exercise of religion and forbids Congress from passing laws respecting its establishment. In a 2002 study, 59% of Americans said that religion played a "very important role in their lives", a far higher figure than that of any other wealthy nation.[158] According to a 2007 survey, 78.4% of adults identified themselves as Christian,[] down from 86.4% in 1990.[]

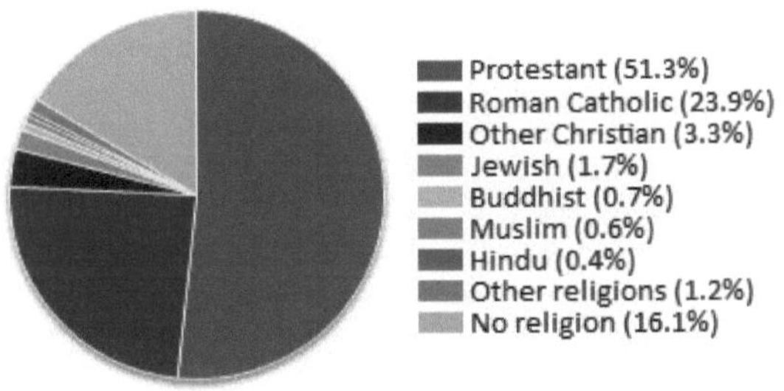


A pie chart of religious groups in the U.S. (2007)

Protestant denominations accounted for 51.3%, while Roman Catholicism, at 23.9%, was the largest individual denomination.[] The total reporting non-Christian religions in 2007 was 4.7%, up from 3.3% in 1990.[] Other religions include Judaism (1.7%), Buddhism (0.7%), Islam (0.6%), Hinduism (0.4%), and Unitarian Universalism

(0.3%).[] The survey also reported that 16.1% of Americans described themselves as agnostic, atheist, or simply having no religion, up from 8.2% in 1990.[] [] []

There are also Baha'i, Sikh, Jain, Shinto, Confucian, Taoist, Druid, Native American, Wiccan, humanist and deist communities.[159] Doubt about the existence of a god or gods is growing rapidly among Americans under 30.[160] Polls show that overall American confidence in organized religion is declining,[161] and that younger Americans in particular are becoming increasingly irreligious.[162]

### Family structure

In 2007, 58% of Americans age 18 and over were married, 6% were widowed, 10% were divorced, and 25% had never been married.[163] Women now work mostly outside the home and receive a majority of bachelor's degrees.[164]

The U.S. teenage pregnancy rate, 79.8 per 1,000 women, is the highest among OECD nations.[165] Between 2007 and 2010, the highest teenage birth rate was in Mississippi, and the lowest in New Hampshire.[166] While the abortion rate is falling, the abortion ratio of 241 per 1,000 live births and abortion rate of 15 per 1,000 women aged 15–44 remain higher than those of most Western nations.[167]

## Government and politics

U.S. Capitol seats the Congress: the Senate, left; the House, right

The White House used by the U.S. President

Supreme Court and offices for nine Justices

The United States is the world's oldest surviving federation. It is a constitutional republic and representative democracy, "in which majority rule is tempered by minority rights protected by law".[168] The government is regulated by a system of checks and balances defined by the U.S. Constitution, which serves as the country's supreme legal document.[169]

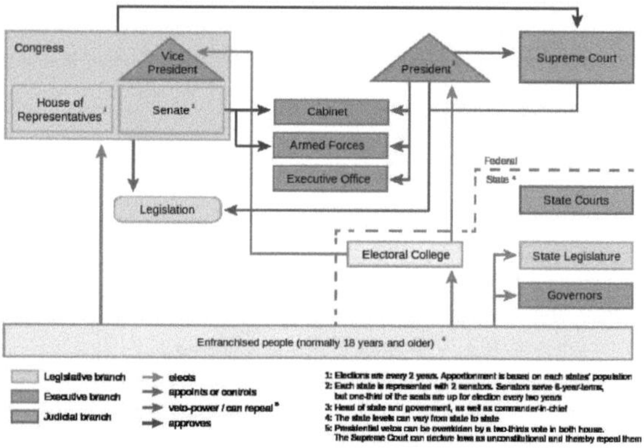


Political system of the United States

In the American federalist system, citizens are usually subject to three levels of government: federal, state, and local. The local government's duties are commonly split between county and municipal governments. In almost all cases, executive and legislative officials are elected by a plurality vote of citizens by district. There is no proportional representation at the federal level, and it is very rare at lower levels.

The federal government is composed of three branches:

- Legislative: The bicameral Congress, made up of the Senate and the House of Representatives, makes federal law, declares war, approves treaties, has the power of the purse,[170] and has the power of impeachment, by which it can remove sitting members of the government.[171]

- Executive: The president is the commander-in-chief of the military, can veto legislative bills before they become law (subject to Congressional override), and appoints the members of the Cabinet (subject to Senate approval) and other officers, who administer and enforce federal laws and policies.[172]
- Judicial: The Supreme Court and lower federal courts, whose judges are appointed by the president with Senate approval, interpret laws and overturn those they find unconstitutional.

The House of Representatives has 435 voting members, each representing a congressional district for a two-year term. House seats are apportioned among the states by population every tenth year. As of the 2010 census, seven states have the minimum of one representative, while California, the most populous state, has 53.[173]

The Senate has 100 members with each state having two senators, elected at-large to six-year terms; one third of Senate seats are up for election every other year. The president serves a four-year term and may be elected to the office no more than twice. The president is not elected by direct vote, but by an indirect electoral college system in which the determining votes are apportioned to the states and the District of Columbia.[174] The Supreme Court, led by the Chief Justice of the United States, has nine members, who serve for life.[175]

The state governments are structured in roughly similar fashion; Nebraska uniquely has a unicameral legislature.[176] The governor (chief executive) of each state is directly elected. Some state judges and cabinet officers are appointed by the governors of the respective states, while others are elected by popular vote.

The original text of the Constitution establishes the structure and responsibilities of the federal government and its relationship with the individual states. Article One protects the right to the "great writ" of habeas corpus, The Constitution has been amended 27 times;[177] the first 10 amendments, which make up the Bill of Rights, and the Fourteenth Amendment form the central basis of Americans' individual rights. All laws and governmental procedures are subject to judicial review and any law ruled in violation of the Constitution is voided. The principle of judicial review, not explicitly mentioned in the Constitution, was declared by the Supreme Court in *Marbury v. Madison* (1803),[178] a decision handed down by Chief Justice John Marshall.[179]

## Political divisions

Further information: Territorial evolution of the United States and United States territorial acquisitions

The United States is a federal union of 50 states. The original 13 states were the successors of the 13 colonies that rebelled against British rule. Early in the country's history, three new states were organized on territory separated from the claims of the existing states: Kentucky from Virginia; Tennessee from North Carolina; and Maine from Massachusetts. Most of the other states have been carved from territories obtained through war or purchase by the U.S. government. One set of exceptions includes Vermont, Texas, and Hawaii: each was an independent republic before joining the union. During the American Civil War, West Virginia broke away from Virginia. The most recent state—Hawaii—achieved statehood on August 21, 1959.[180] The states do not have the right to unilaterally secede from the union.

The states compose the vast bulk of the U.S. land mass; the two other areas considered integral parts of the country are the District of Columbia, the federal district where the capital, Washington, is located; and Palmyra Atoll, an uninhabited but incorporated territory in the Pacific Ocean. The United States also possesses five major overseas territories: Puerto Rico and the United States Virgin Islands in the Caribbean; and American Samoa, Guam, and the Northern Mariana Islands in the Pacific.[181] Those born in the major territories are birthright U.S. citizens except Samoans. Samoans born in American Samoa are born U.S. nationals, and may become naturalized citizens.[182] American citizens residing in the territories have fundamental constitutional protections and elective self-government, with a territorial Member of Congress, but they do not vote for president as states. Territories have personal and business tax regimes different from that of states.[183]

The United States also observes tribal sovereignty of the Native Nations. Though reservations are within state borders, the reservation is a sovereign. While the United States recognizes this sovereignty, other countries may not.[184]

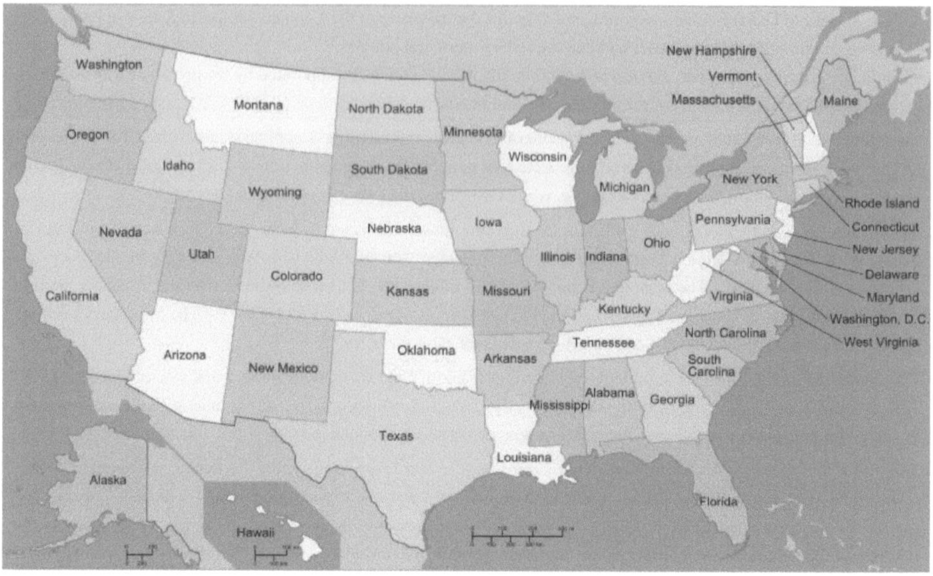


## Parties and elections

(from left to right) House Majority Leader Eric Cantor, House Minority Leader Nancy Pelosi, House Speaker John Boehner, President Barack Obama, Senate Majority Leader Harry Reid, Senate Minority Leader Mitch McConnell at the White House in 2011

The United States has operated under a two-party system for most of its history.[] For elective offices at most levels, state-administered primary elections choose the major party nominees for subsequent general elections. Since the general election of 1856, the major parties have been the Democratic Party, founded in 1824, and the Republican Party, founded in 1854. Since the Civil War, only one third-party presidential candidate—former president Theodore Roosevelt, running as a Progressive in 1912—has won as much as 20% of the popular vote. The third-largest political party is the Libertarian Party.

Within American political culture, the Republican Party is considered center-right or conservative and the Democratic Party is considered center-left or liberal.[185] The states of the Northeast and West Coast and some of the Great Lakes states, known as "blue states", are relatively liberal. The "red states" of the South and parts of the Great Plains and Rocky Mountains are relatively conservative.

The winner of the 2008 presidential election and the 2012 presidential election, Democrat Barack Obama, is the 44th U.S. president.

In the 113th United States Congress, the House of Representatives is controlled by the Republican Party, while the Democratic Party has control of the Senate. The Senate currently consists of 52 Democrats, two independents who caucus with the Democrats, and 46 Republicans; the House consists of 234 Republicans and 201 Democrats.[186] There are 30 Republican and 19 Democratic state governors, as well as one independent.[187]

Since the founding of the United States until 2000s, the country's governance has been primarily dominated by White Anglo-Saxon Protestants (WASPs). However, the situation has changed recently and of the top 17 positions (four national candidates of the two major party in the 2012 U.S. presidential election, four leaders in 112th United States Congress, and nine Supreme Court Justices) there is only one WASP.[188] [189] [190]

## Foreign relations

The United States has established foreign relations. It is a permanent member of the United Nations Security Council, and New York City hosts the United Nations Headquarters. It is a member of the G8,[191] G20, and Organisation for Economic Co-operation and Development. Almost all countries have embassies in Washington, D.C., and many have consulates around the country. Likewise, nearly all nations host American diplomatic missions. However, Cuba, Iran, North Korea, Bhutan, and the Republic of China (Taiwan) do not have formal diplomatic relations with the United States (although the U.S. still supplies Taiwan with military equipment).

British Foreign Secretary William Hague and U.S. Secretary of State Hillary Clinton, May 2010

The United States has a "special relationship" with the United Kingdom[192] and strong ties with Canada,[193] Australia,[194] New Zealand,[195] the Philippines,[196] Japan,[197] South Korea,[198] Israel,[199] and several European countries such as Russia, France and Germany. It works closely with Russia and fellow NATO members on military and security issues and with its neighbors through the Organization of American States and free trade agreements such as the trilateral North American Free Trade Agreement with Canada and Mexico. In 2008, the United States spent a net $25.4 billion on official development assistance, the most in the world. As a share of America's large gross national income (GNI), however, the U.S. contribution of 0.18% ranked last among 22 donor states. By contrast, private overseas giving by Americans is relatively generous.[200]

## Government finance

Taxes are levied in the United States at the federal, state and local government level. These include taxes on income, payroll, property, sales, imports, estates and gifts, as well as various fees. In 2010 taxes collected by federal, state and municipal governments amounted to 24.8% of GDP.[201] During FY2012, the federal government collected approximately $2.45 trillion in tax revenue, up $147 billion or 6% versus FY2011 revenues of $2.30 trillion. Primary receipt categories included individual income taxes ($1,132B or 47%), Social Security/Social Insurance taxes ($845B or 35%), and corporate taxes ($242B or 10%).[]

US taxation is generally progressive, especially the federal income taxes, and is among the most progressive in the developed world.[202] [] [203] [204] [205] In 2009 the top 10% of earners, with 36% of the nation's income, paid 78.2% of the federal personal income tax burden, while the bottom 40% had a negative liability.[] Payroll taxes are less progressive, as the entitlement programs they ostensibly fund have historically not been viewed as welfare transfers.[206] [207] The top 10% paid 51.8% of total federal taxes in 2009, and the top 1%, with 13.4% of pre-tax national income, paid 22.3% of federal taxes.[] In 2013 the Tax Policy Center projected total federal effective tax rates of 35.5% for the top 1%, 27.2% for the top quintile, 13.8% for the middle quintile, and –2.7% for the bottom quintile.[208] [209] State and local taxes vary widely, but are generally less progressive than federal taxes as they rely heavily on broadly borne regressive sales and property taxes that yield less volatile revenue streams, though their consideration doesn't eliminate the progressive nature of overall taxation.[210] [211]

During FY 2012, the federal government spent $3.54 trillion on a budget or cash basis, down $60 billion or 1.7% vs. FY 2011 spending of $3.60 trillion. Major categories of FY 2012 spending included: Medicare & Medicaid ($802B or 23% of spending), Social Security ($768B or 22%), Defense Department ($670B or 19%), non-defense discretionary ($615B or 17%), other mandatory ($461B or 13%) and interest ($223B or 6%).[]

### Public debt

In March 2013, US federal government debt held by the public was approximately $11.888 trillion, or about 75% of US GDP. Intra-governmental holdings stood at $4.861 trillion, giving a combined total debt of $16.749 trillion.[] [] By 2012 total federal debt had surpassed 100% of US GDP.[212] The US has a credit rating of AA+ from Standard & Poor's, AAA from Fitch, and Aaa from Moody's.[213]

U.S. federal debt held by the public as a percentage of GDP, from 1940 to 2012.

Historically, the US public debt as a share of GDP increased during wars and recessions, and subsequently declined. For example, debt held by the public as a share of GDP peaked just after World War II (113% of GDP in 1945), but then fell over the following 30 years. In recent decades, large budget deficits and the resulting increases in debt have led to concern about the long-term sustainability of the federal government's fiscal policies.[214] However, these concerns are not universally shared.[215]

## Military

The president holds the title of commander-in-chief of the nation's armed forces and appoints its leaders, the secretary of defense and the Joint Chiefs of Staff. The United States Department of Defense administers the armed forces, including the Army, Navy, Marine Corps, and Air Force. The Coast Guard is run by the Department of Homeland Security in peacetime and the Department of the Navy in time of war. In 2008, the armed forces had 1.4 million personnel on active duty. The Reserves and National Guard brought the total number of troops to 2.3 million. The Department of Defense also employed about 700,000 civilians, not including contractors.[216]

Military service is voluntary, though conscription may occur in wartime through the Selective Service System.[217] American forces can be rapidly deployed by the Air Force's large fleet of transport aircraft, the Navy's 11 active aircraft carriers, and Marine Expeditionary Units at sea with the Navy's Atlantic and Pacific fleets. The military operates 865 bases and facilities abroad,[218] and maintains deployments greater than 100 active duty personnel in 25 foreign countries.[219] The extent of this global military presence has prompted some scholars to describe the United States as maintaining an "empire of bases".[220] The United States is in many ways a new kind of entity in the geopolitical bestiary: a destabilizing hegemon, a revolutionary empire.[]

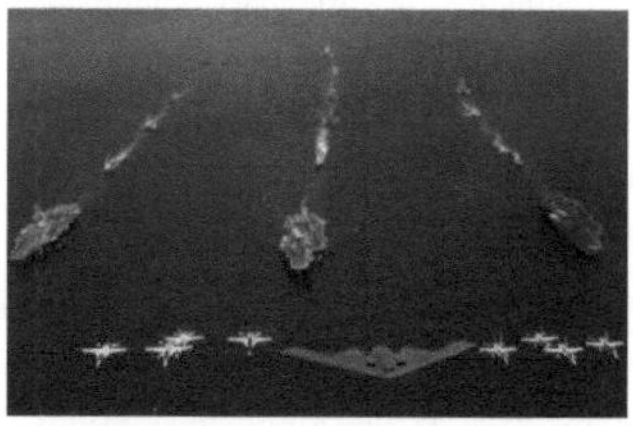

The carrier strike groups of the *Kitty Hawk*, *Ronald Reagan*, and *Abraham Lincoln* with aircraft from the Marine Corps, Navy, and Air Force.

Total U.S. military spending in 2011, more than $700 billion, was 41% of global military spending and equal to the next 14 largest national military expenditures combined. At 4.7% of GDP, the rate was the second-highest among the top 15 military spenders, after Saudi Arabia.[221] U.S. defense spending as a percentage of GDP ranks 23rd globally as of 2012 according to the CIA.[222] Defense's share of U.S. spending has generally declined in recent decades, from

Cold War peaks of 14.2% of GDP in 1953 and 69.5% of federal outlays in 1954 to 4.7% of GDP and 18.8% of federal outlays in 2011.[223]

The proposed base Department of Defense budget for 2012, $553 billion, is a 4.2% increase over 2011; an additional $118 billion is proposed for the military campaigns in Iraq and Afghanistan.[224] The last American troops serving in Iraq departed in December 2011;[225] 4,484 servicemen were killed during the Iraq War.[226] Approximately 90,000 U.S. troops were serving in Afghanistan as of April 2012;[] as of May 21, 2013, 2,039 had been killed during the War in Afghanistan.[227]

## Law enforcement

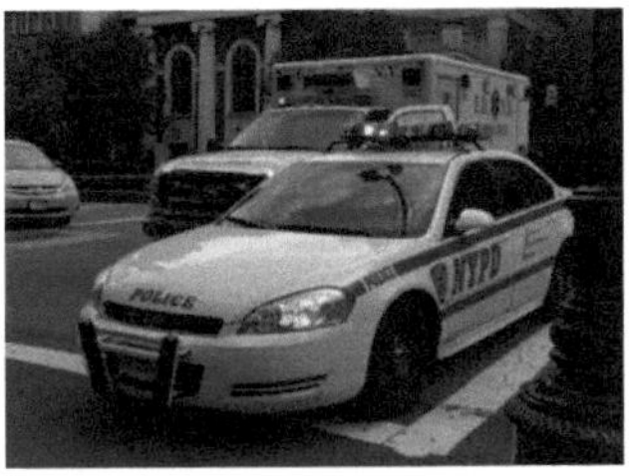

Law enforcement in the U.S. is maintained primarily by local police departments. The New York City Police Department (NYPD) is the largest in the country.[228]

Law enforcement in the United States is primarily the responsibility of local police and sheriff's departments, with state police providing broader services. Federal agencies such as the Federal Bureau of Investigation (FBI) and the U.S. Marshals Service have specialized duties.[229] At the federal level and in almost every state, jurisprudence operates on a common law system.

State courts conduct most criminal trials; federal courts handle certain designated crimes as well as certain appeals from the state criminal courts. Federal law prohibits a variety of drugs, although states sometimes pass laws in conflict with federal regulations. The smoking age is generally 18, and the drinking age is generally 21.

Among developed nations, the United States has above-average levels of violent crime and particularly high levels of gun violence and homicide.[230] There were 4.7 murders per 100,000 persons in 2011, 14.5% fewer than in 2000 (5.5), and 19.0% fewer since a recent peak of 5.8 in 2006.[] [] Gun ownership rights are the subject of contentious political debate.

The United States has the highest documented incarceration rate[] and total prison population[231] in the world. At the start of 2008, more than 2.3 million people were incarcerated, more than one in every 100 adults.[232] The current rate is about seven times the 1980 figure,[233] and over three times the figure in Poland, the Organisation for Economic Co-operation and Development (OECD) country with the next highest rate.[234] African American males are jailed at about six times the rate of white males and three times the rate of Hispanic males.[] The country's high rate of incarceration is largely due to sentencing and drug policies.[] []

Capital punishment is sanctioned in the United States for certain federal and military crimes, and used in 32 states.[235] While there are 32 states which include capital punishment within their sentencing statutes, some states (such as New Hampshire and Kansas) have yet to execute anyone since 1976, as demonstrated by the lack of any executions by these states out of the 1317 total executions which have taken place as of December 5, 2012 [236] No executions took place from 1967 to 1977, owing in part to a U.S. Supreme Court ruling striking down arbitrary imposition of the death penalty. In 1976, that Court ruled that, under appropriate circumstances, capital punishment may constitutionally be imposed; since the decision there have been more than 1,300 executions, a majority of these taking place in three states: Texas, Virginia, and Oklahoma.[237] Five state legislatures in the modern era have abolished the death penalty, though two of those laws (in New Mexico and Connecticut) were not retroactive. Additionally, state courts in Massachusetts and New York struck down death penalty statutes and their legislatures took no action in response. In 2010, the country had the fifth highest number of executions in the world, following China, Iran, North Korea, and Yemen.[238]

## Economy

| Economic indicators | | |
|---|---|---|
| Nominal GDP | 15.984 trillion (Q1 2013) | [] |
| Real GDP growth | 1.8% (Q1 2013, annualized) | |
| | 2.2% (2012) | [239] |
| CPI inflation | 2.0% (February 2012 – February 2013) | [240] |
| Employment-to-population ratio | 58.5% (March 2013) | [241] |
| Unemployment | 7.4% (July 2013) | [242] |
| Labor force participation rate | 63.3% (March 2013) | [243] |
| Poverty | 15.1% (2010) | [] |
| Public debt | $16.433 trillion (Q4 2012) | [244] |
| Household net worth | $58.5 trillion (Q4 2011) | [245] |

The United States has a capitalist mixed economy, which is fueled by abundant natural resources, a well-developed infrastructure, and high productivity.[246] According to the International Monetary Fund, the U.S. GDP of $15.1 trillion constitutes 22% of the gross world product at market exchange rates and over 19% of the gross world product at purchasing power parity (PPP).[] Though larger than any other nations, its national GDP was about 5% smaller at PPP in 2011 than the European Union's, whose population is around 62% higher.[247] The country ranks ninth in the world in nominal GDP per capita and sixth in GDP per capita at PPP.[] The U.S. dollar is the world's primary reserve currency.[248]

The United States is the largest importer of goods and second largest exporter, though exports per capita are relatively low. In 2010, the total U.S. trade deficit was $635 billion.[] Canada, China, Mexico, Japan, and Germany are its top trading partners.[249] In 2010, oil was the largest import commodity, while transportation equipment was the country's largest export.[] China is the largest foreign holder of U.S. public debt.[250]

In 2009, the private sector was estimated to constitute 86.4% of the economy, with federal government activity accounting for 4.3% and state and local government activity (including federal transfers) the remaining 9.3%.[252] While its economy has reached a postindustrial level of development and its service sector constitutes 67.8% of GDP, the United States remains an industrial power.[] The leading business field by gross business receipts is wholesale and retail trade; by net income it is manufacturing.[253]

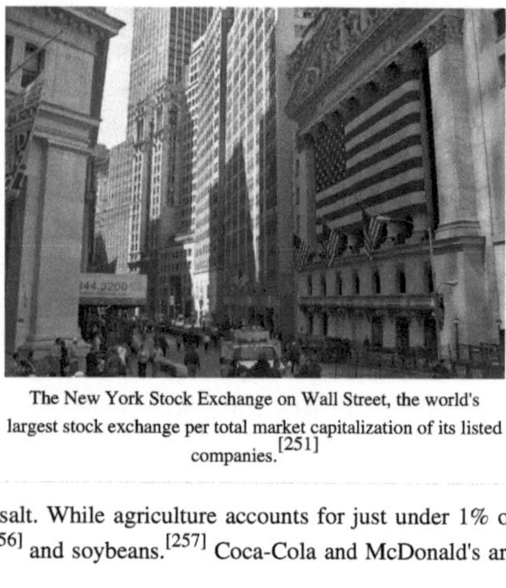

The New York Stock Exchange on Wall Street, the world's largest stock exchange per total market capitalization of its listed companies.[251]

Chemical products are the leading manufacturing field.[254] The United States is the third largest producer of oil in the world, as well as its largest importer.[255] It is the world's number one producer of electrical and nuclear energy, as well as liquid natural gas, sulfur, phosphates, and salt. While agriculture accounts for just under 1% of GDP,[] the United States is the world's top producer of corn[256] and soybeans.[257] Coca-Cola and McDonald's are the two most recognized brands in the world.[258]

Consumer spending comprises 71% of the US economy in 2013.[259] In August 2010, the American labor force consisted of 154.1 million people. With 21.2 million people, government is the leading field of employment. The largest private employment sector is health care and social assistance, with 16.4 million people. About 12% of workers are unionized, compared to 30% in Western Europe.[260] The World Bank ranks the United States first in the ease of hiring and firing workers.[] In 2009, the United States had the third highest labor productivity per person in the world, behind Luxembourg and Norway. It was fourth in productivity per hour, behind those two countries and the Netherlands.[261]

The 2008-2012 global recession had a significant impact on the United States. For example, persistent high unemployment remains, along with low consumer confidence, the continuing decline in home values and increase in foreclosures and personal bankruptcies, an escalating federal debt crisis, inflation, and rising petroleum and food prices. In fact, a 2011 poll found that more than half of all Americans think the U.S. is still in recession or even depression, despite official data that shows a historically modest recovery.[262]

## Income, poverty, and wealth

Americans have the highest average household and employee income among OECD nations, and, as of 2007, the second highest median household income, behind only Luxembourg.[] [] According to the Census Bureau real median household income was $50,502 in 2011, down from $51,144 in 2010.[263] The Global Food Security Index ranked the US #1 in food affordability and overall food security in March 2013.[264] On average Americans have over twice as much living space per dwelling and per person as Europeans, and more than every European nation.[]

The US economy is currently embroiled in the economic downturn which followed the Financial crisis of 2007–2008, with output still below potential according to the CBO[265] and unemployment still above historic trends.[266] As of February 2013, the unemployment rate was 7.7% or 12.0 million people, while the government's broader U-6 unemployment rate, which includes the part-time underemployed was 14.3% or 22.2 million. With a record proportion of long term unemployed, continued decreasing household income, tax hikes, and new federal budget cuts, the US economy remained in a jobless recovery.[] [267] Half of the U.S. population lives in poverty or is low-income, according to U.S. census data.[268] According to a survey by the Associated Press, four out of five U.S. adults struggle with joblessness, near-poverty or reliance on welfare for at least parts of their lives.[269]

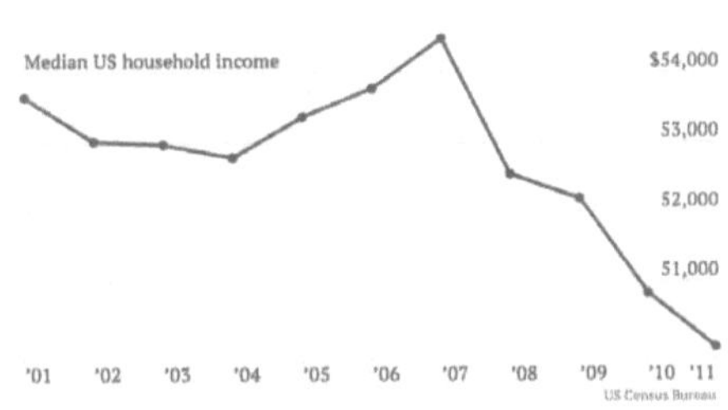


The median US household income, 2001–2011.

While inflation-adjusted ("real") household income had been increasing almost every year from 1947 to 1999, it has since been flat and even decreased recently.[270] Extreme poverty in the United States, meaning households living on less than $2 per day before government benefits, doubled from 1996 levels to 1.5 million households in 2011, including 2.8 million children.[271] In 2011 16.7 million children lived in food insecure households, about 35% more than 2007 levels, though only 1.1% of US children, or 845,000, saw reduced food intake or disrupted eating patterns at some point during the year, and most cases weren't chronic.[272]

There were about 643,000 sheltered and unsheltered homeless persons in the U.S. in January 2009. Almost two-thirds stayed in an emergency shelter or transitional housing program and the other third were living on the street, in an abandoned building, or another place not meant for human habitation. About 1.56 million people, or about 0.5% of the U.S. population, used an emergency shelter or a transitional housing program between October 1, 2008 and September 30, 2009.[] The U.S. welfare state is one of the least extensive in the developed world, reducing both relative poverty and absolute poverty by considerably less than the mean for rich nations,[] [273] [274] though combined private and public social expenditures per capita are relatively high and Americans face much lower consumption taxes than poor Europeans.[275]

While the American welfare state effectively reduces poverty among the elderly,[276] it provides relatively little assistance to the young.[277] A 2007 UNICEF study of children's well-being in 21 industrialized nations, based on factors like income relative to each nation's own median, self-reported risky behavior, and family relationship quality, ranked the United States next to last.[278] The share of income of the top 1%—21.8% of total reported income in 2005—has more than doubled since 1980,[279] leaving the United States with the greatest income inequality among developed nations.[] [280]

Poverty in the US has been increasing as median incomes have declined. Analyses using a common data set for comparisons tend to find that the US has a lower absolute poverty rate by market income than most other wealthy nations.[281] Over 80% of poor American households have air conditioning, three quarters own at least one automobile, about 40% own their homes, and the average poor American has more living space than the general population average in every European nation except Luxembourg and Denmark. Most of them have a refrigerator, stove, microwave, telephone, and television. About half have computers and less than half have internet service.[]

A middle-class suburban housing development in San Jose, California

Wealth, like income and taxes, is highly concentrated: The richest 10% of the adult population possesses 69.8% of the country's household wealth, the second-highest share among developed nations.[282] In 2013 the United Nations Development Programme ranked the United States 16th among 132 countries on its inequality-adjusted human development index (IHDI), 13 places lower than in the standard HDI.[283] For the year 2012, the United States ranks 12th on the Legatum Prosperity Index[284] and 19th on the Corruption Perceptions Index.[285]

Between June 2007 and November 2008 the global recession led to falling asset prices around the world. Assets owned by Americans lost about a quarter of their value.[286] Since peaking in the second quarter of 2007, household wealth is down $14 trillion.[287] At the end of 2008, household debt amounted to $13.8 trillion.[288] By some measures, the US has more millionaires per capita than any other nation, ranks in the top 14 in billionaires per capita,[289] and has more billionaires and millionaires than any other nation and all of Europe; most of them self-made.[290] [291] [292]

# Infrastructure

## Transportation

Personal transportation is dominated by automobiles, which operate on a network of 13 million roads, including one of the world's longest highway systems.[294] The world's second largest automobile market,[295] the United States has the highest rate of per-capita vehicle ownership in the world, with 765 vehicles per 1,000 Americans.[296] About 40% of personal vehicles are vans, SUVs, or light trucks.[297] The average American adult (accounting for all drivers and non-drivers) spends 55 minutes driving every day, traveling 29 miles ( km).[298]

The Interstate Highway System, which extends 46876 miles ( km)[293]

Mass transit accounts for 9% of total U.S. work trips.[299] [300] While transport of goods by rail is extensive, relatively few people use rail to travel,[301] though ridership on Amtrak, the national intercity passenger rail system, grew by almost 37% between 2000 and 2010.[302] Also, light rail development has increased in recent years.[303] Bicycle usage for work commutes is minimal.[304]

The civil airline industry is entirely privately owned and has been largely deregulated since 1978, while most major airports are publicly owned. The three largest airlines in the world by passengers carried are U.S.-based; American Airlines is number one after its 2013 acquisition of US Airways.[305] Of the world's 30 busiest passenger airports, 16 are in the United States, including the busiest, Hartsfield-Jackson Atlanta International Airport.[306]

Further information: List of airlines of the United States

## Energy

The United States energy market is 29,000 terawatt hours per year. Energy consumption per capita is 7.8 tons of oil equivalent per year, the 10th highest rate in the world. In 2005, 40% of this energy came from petroleum, 23% from coal, and 22% from natural gas. The remainder was supplied by nuclear power and renewable energy sources.[307] The United States is the world's largest consumer of petroleum.[308]

For decades, nuclear power has played a limited role relative to many other developed countries, in part because of public perception in the wake of a 1979 accident. In 2007, several applications for new nuclear plants were filed.[309] The United States has 27% of global coal reserves.[] It is the world's largest producer of natural gas and crude oil.[310]

## Science and technology

The United States has been a leader in scientific research and technological innovation since the late 19th century. In 1876, Alexander Graham Bell was awarded the first U.S. patent for the telephone. Thomas Edison's laboratory developed the phonograph, the first long-lasting light bulb, and the first viable movie camera.[311] Nikola Tesla pioneered alternating current, the AC motor, and radio. In the early 20th century, the automobile companies of Ransom E. Olds and Henry Ford popularized the assembly line. The Wright brothers, in 1903, made the first sustained and controlled heavier-than-air powered flight.[312]

Neil Armstrong was the first person to walk on the Moon.

The rise of Nazism in the 1930s led many European scientists, including Albert Einstein, Enrico Fermi, and John von Neumann, to immigrate to the United States. During World War II, the Manhattan Project developed nuclear weapons, ushering in the Atomic Age. The Space Race produced rapid advances in rocketry, materials science, and computers. IBM, Apple Computer, and Microsoft refined and popularized the personal computer.

The United States largely developed the ARPANET and its successor, the Internet. Today, 64% of research and development funding comes from the private sector.[313] The United States leads the world in scientific research papers and impact factor.[314] As of April 2010, 77% of American households owned at least one computer, and 68% had broadband Internet service.[315] 85% of Americans also own a mobile phone as of 2011.[316] The country is the primary developer and grower of genetically modified food, representing half of the world's biotech crops.[317]

## Education

American public education is operated by state and local governments, regulated by the United States Department of Education through restrictions on federal grants. In most states, children are required to attend school from the age of six or seven (generally, kindergarten or first grade) until they turn 18 (generally bringing them through twelfth grade, the end of high school); some states allow students to leave school at 16 or 17.[319] About 12% of children are enrolled in parochial or nonsectarian private schools. Just over 2% of children are homeschooled.[320]

Some 80% of U.S. college students attend public universities such as the University of Virginia, founded by Thomas Jefferson[318]

The United States has many competitive private and public institutions of higher education. According to prominent international rankings, 13 or 15 American colleges and universities are ranked among the top 20 in the world.[321] [322] There are also local community colleges with generally more open admission policies, shorter academic programs, and lower tuition. Of Americans 25 and older, 84.6% graduated from high school, 52.6% attended some college, 27.2% earned a bachelor's degree, and 9.6% earned graduate degrees.[323] The basic literacy rate is approximately 99%.[] [324] The United Nations assigns the United States an Education Index of 0.97, tying it for 12th in the world.[325]

## Health

The United States has life expectancy of 78.4 years at birth, up from 75.2 years in 1990, ranks it 50th among 221 nations, and 27th out of the 34 industrialized OECD countries, down from 20th in 1990.[326] [] Increasing obesity in the United States and health improvements elsewhere have contributed to lowering the country's rank in life expectancy from 1987, when it was 11th in the world.[327] Obesity rates in the United States are among the highest in the world.[328] Approximately one-third of the adult population is obese and an additional third is overweight;[329] the obesity rate, the highest in the industrialized world, has more than doubled in the last quarter-century.[330] Obesity-related type 2 diabetes is considered epidemic by health care professionals.[331] The infant mortality rate of 6.06 per thousand places the United States 176th highest out of 222 countries.[332]

The Texas Medical Center in Houston is the world's largest medical center.

In 2010, coronary artery disease, lung cancer, stroke, chronic obstructive pulmonary diseases, and traffic accidents caused the most years of life lost in the US. Low back pain, depression, musculoskeletal disorders, neck pain, and anxiety caused the most years lost to disability. The most deleterious risk factors were poor diet, tobacco smoking, obesity, high blood pressure, high blood sugar, physical inactivity, and alcohol use. Alzheimer's disease, drug abuse, kidney disease and cancer, and falls caused the most additional years of life lost over their age-adjusted 1990 per-capita rates.[]

The U.S. is a global leader in medical innovation. America solely developed or contributed significantly to 9 of the top 10 most important medical innovations since 1975 as ranked by a 2001 poll of physicians, while the EU and Switzerland together contributed to five. Since 1966 Americans have received more Nobel Prizes in Medicine than the rest of the world combined. From 1989 to 2002 four times more money was invested in private biotechnology companies in America than in Europe.[333] [334]

The U.S. health-care system far outspends any other nations, measured in both per capita spending and percentage of GDP.[335] In 2008, the US spent more on health care per capita ($7,146), and as percentage of GDP (15.2%), than any other nation. Health-care coverage in the United States is a combination of public and private efforts, and is not universal as in all other developed countries. In 2004, private insurance paid for 36% of personal health expenditures, private out-of-pocket payments covered 15%, and federal, state, and local governments paid for 44%.[]

In 2010, 49.9 million residents or 16.3% of the population did not carry health insurance. The main cause of this rise is the drop in the number of Americans with employer-sponsored health insurance.[] The subject of uninsured and underinsured Americans is a major political issue.[336] [337] In 2006, Massachusetts became the first state to mandate universal health insurance.[338] In 2007, 62.1% of filers for bankruptcy blamed medical expenses. About 25% of all senior citizens declare bankruptcy because of medical expenses, and 43% are forced to mortgage or sell their primary residence.[339] Federal legislation passed in early 2010 would ostensibly create a near-universal health insurance system around the country by 2014, though the bill and its ultimate impact are issues of controversy.[340] [341]

## Culture

The United States is a multicultural nation, home to a wide variety of ethnic groups, traditions, and values.[8] [343] Aside from the relatively small Native American and Native Hawaiian populations, nearly all Americans or their ancestors settled or immigrated within the past five centuries.[344] Mainstream American culture is a Western culture largely derived from the traditions of European immigrants with influences from many other sources, such as traditions brought by slaves from Africa.[8] [345] More recent immigration from Asia and especially Latin America has added to a cultural mix that has been described as both a homogenizing melting pot, and a heterogeneous salad bowl in which immigrants and their descendants retain distinctive cultural characteristics.[8]

The Statue of Liberty in New York City is a symbol of both the U.S. and ideals of freedom, democracy, and opportunity.[342]

American culture is considered the most individualistic in the world.[346] The American Dream, or the perception that Americans enjoy high social mobility, plays a key role in attracting immigrants.[347] By some measures, social mobility is actually lower than other high-income countries such as France, Germany, Canada, Australia, and the Nordic countries,[348] though such studies are based on relative comparisons within each nation rather than absolute wealth earned throughout one's life, the US having both a more stretched-out income distribution and a higher median income than those nations.[347] [349] While the mainstream culture holds that the United States is a classless society,[350] scholars identify significant differences between the country's social classes, affecting socialization, language, and values.[351]

The American middle and professional class has initiated many contemporary social trends such as modern feminism, environmentalism, and multiculturalism.[352] Americans' self-images, social viewpoints, and cultural expectations are associated with their occupations to an unusually close degree.[353] While Americans tend greatly to value socioeconomic achievement, being ordinary or average is generally seen as a positive attribute.[354]

### Popular media

The world's first commercial motion picture exhibition was given in New York City in 1894, using Thomas Edison's Kinetoscope. The next year saw the first commercial screening of a projected film, also in New York, and the United States was in the forefront of sound film's development in the following decades. Since the early 20th century, the U.S. film industry has largely been based in and around Hollywood, California.

Director D. W. Griffith was central to the development of film grammar and Orson Welles's *Citizen Kane* (1941) is frequently cited as the greatest film of all time.[355] [356] American screen actors like John Wayne and Marilyn Monroe have become iconic figures, while producer/entrepreneur Walt Disney was a leader in both animated film and movie merchandising. Hollywood is also one of the leaders in motion picture production.[357]

Americans are the heaviest television viewers in the world,[358] and the average viewing time continues to rise, reaching five hours a day in 2006.[359] The four major broadcast television networks are all commercial entities. Americans listen to radio programming, also largely commercialized, on average just over two-and-a-half hours a day.[360] Aside from web portals and search engines, the most popular websites are Facebook, YouTube, Wikipedia, Blogger, eBay, and Craigslist.[]

The rhythmic and lyrical styles of African-American music have deeply influenced American music at large, distinguishing it from European traditions. Elements from folk idioms such as the blues and what is now known as old-time music were adopted and transformed into popular genres with global audiences. Jazz was developed by innovators such as Louis Armstrong and Duke Ellington early in the 20th century. Country music developed in the 1920s, and rhythm and blues in the 1940s.[361]

Elvis Presley and Chuck Berry were among the mid-1950s pioneers of rock and roll. In the 1960s, Bob Dylan emerged from the folk revival to become one of America's most celebrated songwriters and James Brown led the development of funk. More recent American creations include hip hop and house music. American pop stars such as Presley, Michael Jackson, and Madonna have become global celebrities.[361]

## Literature, philosophy, and the arts

In the 18th and early 19th centuries, American art and literature took most of its cues from Europe. Writers such as Nathaniel Hawthorne, Edgar Allan Poe, and Henry David Thoreau established a distinctive American literary voice by the middle of the 19th century. Mark Twain and poet Walt Whitman were major figures in the century's second half; Emily Dickinson, virtually unknown during her lifetime, is now recognized as an essential American poet.[362] A work seen as capturing fundamental aspects of the national experience and character—such as Herman Melville's *Moby-Dick* (1851), Twain's *The Adventures of Huckleberry Finn* (1885), and F. Scott Fitzgerald's *The Great Gatsby* (1925)—may be dubbed the "Great American Novel".[363]

Mark Twain, famous American author and humorist.

Eleven U.S. citizens have won the Nobel Prize in Literature, most recently Toni Morrison in 1993. William Faulkner and Ernest Hemingway are often named among the most influential writers of the 20th century.[364] Popular literary genres such as the Western and hardboiled crime fiction developed in the United States. The Beat Generation writers opened up new literary approaches, as have postmodernist authors such as John Barth, Thomas Pynchon, and Don DeLillo.

The transcendentalists, led by Thoreau and Ralph Waldo Emerson, established the first major American philosophical movement. After the Civil War, Charles Sanders Peirce and then William James and John Dewey were leaders in the development of pragmatism. In the 20th century, the work of W. V. O. Quine and Richard Rorty, and later Noam Chomsky, brought analytic philosophy to the fore of American philosophical academia. John Rawls and Robert Nozick led a revival of political philosophy. Cornel West and Judith Butler have led a continental tradition in American philosophical academia. Globally influential Chicago school economists like Milton Friedman, James M. Buchanan, and Thomas Sowell have transcended discipline to impact various fields in social and political philosophy.[365] [366]

In the visual arts, the Hudson River School was a mid-19th-century movement in the tradition of European naturalism. The realist paintings of Thomas Eakins are now widely celebrated. The 1913 Armory Show in New York City, an exhibition of European modernist art, shocked the public and transformed the U.S. art scene.[367] Georgia O'Keeffe, Marsden Hartley, and others experimented with new, individualistic styles. Major artistic movements such as the abstract expressionism of Jackson Pollock and Willem de Kooning and the pop art of Andy Warhol and Roy Lichtenstein developed largely in the United States. The tide of modernism and then postmodernism has brought fame to American architects such as Frank Lloyd Wright, Philip Johnson, and Frank Gehry.

One of the first major promoters of American theater was impresario P. T. Barnum, who began operating a lower Manhattan entertainment complex in 1841. The team of Harrigan and Hart produced a series of popular musical comedies in New York starting in the late 1870s. In the 20th century, the modern musical form emerged on Broadway; the songs of musical theater composers such as Irving Berlin, Cole Porter, and Stephen Sondheim have become pop standards. Playwright Eugene O'Neill won the Nobel literature prize in 1936; other acclaimed U.S. dramatists include multiple Pulitzer Prize winners Tennessee Williams, Edward Albee, and August Wilson.

Times Square in New York City, the hub of the Broadway Theater District.

Though little known at the time, Charles Ives's work of the 1910s established him as the first major U.S. composer in the classical tradition, while experimentalists such as Henry Cowell and John Cage created a distinctive American approach to classical composition. Aaron Copland and George Gershwin developed a new synthesis of popular and classical music. Choreographers Isadora Duncan and Martha Graham helped create modern dance, while George Balanchine and Jerome Robbins were leaders in 20th-century ballet. Americans have long been important in the modern artistic medium of photography, with major photographers including Alfred Stieglitz, Edward Steichen, and Ansel Adams. The newspaper comic strip and the comic book are both U.S. innovations. Superman, the quintessential comic book superhero, has become an American icon.[368]

## Food

Mainstream American cuisine is similar to that in other Western countries. Wheat is the primary cereal grain. Traditional American cuisine uses indigenous ingredients, such as turkey, venison, potatoes, sweet potatoes, corn, squash, and maple syrup, which were consumed by Native Americans and early European settlers.

Slow-cooked pork and beef barbecue, crab cakes, potato chips, and chocolate chip cookies are distinctively American foods. Soul food, developed by African slaves, is popular around the South and among many African Americans elsewhere. Syncretic cuisines such as Louisiana Creole, Cajun, and Tex-Mex are regionally important.

An apple pie, surrounded by some other icons of American culture.

Characteristic dishes such as apple pie, fried chicken, pizza, hamburgers, and hot dogs derive from the recipes of various immigrants. French fries, Mexican dishes such as burritos and tacos, and pasta dishes freely adapted from Italian sources are widely consumed.[] Americans generally prefer coffee to tea. Marketing by U.S. industries is largely responsible for making orange juice and milk ubiquitous breakfast beverages.[369] [370]

The American fast food industry, the world's largest, pioneered the drive-through format in the 1930s. Fast food consumption has sparked health concerns. During the 1980s and 1990s, Americans' caloric intake rose 24%;[] frequent dining at fast food outlets is associated with what public health officials call the American "obesity epidemic".[371] Highly sweetened soft drinks are widely popular, and sugared beverages account for nine percent of American caloric intake.[372]

## Sports

Swimmer Michael Phelps is the most decorated Olympic athlete of all time.

The market for professional sports in the United States is roughly $69 billion, roughly 50% larger than that of all of Europe, the Middle East, and Africa combined.[373] Baseball has been regarded as the national sport since the late 19th century, while American football is now by several measures the most popular spectator sport.[374] Basketball and ice hockey are the country's next two leading professional team sports. These four major sports, when played professionally, each occupy a season at different, but overlapping, times of the year. College football and basketball attract large audiences. Boxing and horse racing were once the most watched individual sports,[375] but they have been eclipsed by golf and auto racing, particularly NASCAR. In the 21st century, televised mixed martial arts has also gained a strong following of regular viewers.[376] While soccer is much less popular in the United States than in many other nations, it is played widely at the youth and amateur levels; even so, the men's team has been to the past six World Cups and the women are #1 in the women's world rankings. Tennis and many outdoor sports are popular as well.

While most major U.S. sports have evolved out of European practices, basketball, volleyball, skateboarding, snowboarding, and cheerleading are American inventions, which have become popular in other countries. Lacrosse and surfing arose from Native American and Native Hawaiian activities that predate Western contact. Eight Olympic Games have taken place in the United States. The United States has won 2,301 medals at the Summer Olympic Games, more than any other country,[377] and 253 in the Winter Olympic Games, the second most by 2006.[378]

# See also

- Index of United States-related articles
- Outline of the United States
- List of United States cities by population
- List of metropolitan areas of the United States
- List of official United States national symbols
- United States at Wikipedia books

# References

[1] 36 U.S.C. § 302 (http://www.law.cornell.edu/uscode/36/302.html) *National motto*

[2] Simonson, 2010

[3] Dept. of Treasury, 2011

[4] (figure updated automatically).

[8] Adams, J.Q.; Strother-Adams, Pearlie (2001). *Dealing with Diversity*. Chicago: Kendall/Hunt. ISBN 0-7872-8145-X.

[9] Greene, Jack P.; Pole, J.R., eds. (2008). *A Companion to the American Revolution*. pp. 352–361.

[11]

[13] The European Union has a larger collective economy, but is not a single nation.

[15] (http://www.infrastructurereportcard.org/,2009)

[19] Cohen, 2004:History and the Hyperpower

[20] BBC, April 2008:Country Profile: United States of America

[22] DeLear, Byron (July 4, 2013) Who coined 'United States of America'? Mystery might have intriguing answer. (http://www.csmonitor.com/USA/Politics/2013/0704/Who-coined-United-States-of-America-Mystery-might-have-intriguing-answer) "Historians have long tried to pinpoint exactly when the name 'United States of America' was first used and by whom. A new find suggests the man might have been George Washington himself." *Christian Science Monitor* (Boston, MA).

[23] *"To the inhabitants of Virginia," by A PLANTER.* Dixon and Hunter's Virginia Gazette #1287 (http://research.history.org/DigitalLibrary/VirginiaGazette/VGIssueThumbs.cfm?IssueIDNo=76.DH.16) – April 6, 1776, Williamsburg, Virginia. Letter is also included in Peter Force's *American Archives* Vol. 5

[24] Carter, Rusty (August 18, 2012). "You read it here first" (http://www.vagazette.com/news/va-vg-usa-first-0818-20120818,0,4983868.story). *Virginia Gazette*. "He did a search of the archives and found the letter on the front page of the April 6, 1776, edition, published by Hunter & Dixon."

[25] DeLear, Byron (August 16, 2012). "Who coined the name 'United States of America'? Mystery gets new twist." (http://www.csmonitor.com/USA/Politics/The-Vote/2012/0816/Who-coined-the-name-United-States-of-America-Mystery-gets-new-twist#disqus_thread) *Christian Science Monitor* (Boston, MA).

[26] Jefferson's "original Rough draught" of the Declaration of Independence (http://www.princeton.edu/~tjpapers/declaration/declaration.html)

[30] Wilson, Kenneth G. (1993). *The Columbia Guide to Standard American English*. New York: Columbia University Press, pp. 27–28. ISBN 0-231-06989-8.

[31] G. H. Emerson, The Universalist Quarterly and General Review, Vol. 28 (Jan. 1891), p. 49, quoted in Zimmer paper above.

[32] For example, the U.S. embassy in Spain calls itself the embassy of the "Estados Unidos", literally the words "states" and "united", and also uses the initials "EE.UU.", the doubled letters implying plural use in Spanish (http://spanish.madrid.usembassy.gov/) Elsewhere on the site "Estados Unidos de América" is used (http://spanish.madrid.usembassy.gov/es/educacion/benjamin_2013.html)

[33] Smithsonian, 2004, Human Origins

[34] " *The Cambridge encyclopedia of human paleopathology* (http://books.google.com/books?id=qubTdDk1H3IC&pg=PA205)". Arthur C. Aufderheide, Conrado Rodríguez-Martín, Odin Langsjoen (1998). Cambridge University Press. p.205. ISBN 0-521-55203-6

[35] Bianchine, Russo, 1992 pp.225-232

[36] Mann, 2005 p.44

[37] Thornton, 1987 p.49

[38] Kessel, 2005 pp.142-143

[39] Mercer Country Historical Society, 2005

[40] Juergens, 2011, p.69

[41] Ripper, 2008 p.6

[42] Ripper, 2008 p.5

[43] Calloway, 1998, p.55

[44] Vaughan, 1999, p.12

[45] Ranlet, 1999, pp.140-141

[46] Ranlet, 1999, p.137

[47] Rausch, 1994, p.59

[48] Sturm, Circe. "Blood Politics, Racial Classification, and Cherokee National Identity: The Trials and Tribulations of the Cherokee Freedmen", *American Indian Quarterly*, Vol. 22, No. 1/2. (Winter – Spring, 1998), p.231.

[49] Taylor, pp. 33-34

[50] Taylor, pp. 72, 74

[51] Walton, 2009, pp. 29-31

[52] Remini, 2007, pp. 2,3,

[53] Johnson, 1997, pp. 26-30

[54] Berkin, 2007, p. 75

[55] Price, 2003

[56] Vaughan, 1999

[57] Walton, 2009, chapter 3

[58] Lemon, 1987

[59] Gordon, 2004, intro, chapters 1 and 2

[60] Russell, David Lee (2005). *The American Revolution in the Southern Colonies*. Jefferson, N.C., and London: McFarland, p. 12. ISBN 0-7864-0783-2.

[61] Clingan, 2000, p.13

[62] Tadman, 2000, p.1534

[63] Schneider, 2007, p.484

[64] Lien, 1913, p.522

[65] Davis, 1996, p.7

[66] Quirk, 2011, p.195

[67] Gold, 2006, pp.32-35

[68] Walton, 2009, pp. 38,39

[69] Walton, 2009, p. 35

[70] Greene and Pole, *A Companion to the American Revolution* p 357
Jonathan R. Dull, *A Diplomatic History of the American Revolution* (1987) p. 161

Lawrence S. Kaplan, "The Treaty of Paris, 1783: A Historiographical Challenge," *International History Review*, Sept 1983, Vol. 5 Issue 3, pp 431–442

[71] Boyer, 2007, pp.192-193

[72] Walton, 2009, p. 43

[73] Gordon, 2004, pp. 27,29

[75] Smith (2001), *Grant*, pp. 525–526

[76] Carlisle (2008), *The Civil War and Reconstruction*, p. 1

[77] Page 7 lists a total slave population of 3,953,760.

[78] De Rosa, Marshall L. (1997). *The Politics of Dissolution: The Quest for a National Identity and the American Civil War*. Edison, NJ: Transaction. p. 266. ISBN 1-56000-349-9.

[79] Vinovskis, Maris (1990). " *Toward a social history of the American Civil War: exploratory essays* (http://books.google.com/books?id=gySktxKYPGoC&pg=PA6)". Cambridge University Press. p. 6. ISBN 0-521-39559-3.

[80] Brands (2012), *Grant Takes On The Klan*, **American History**, p. 46

[83] Zinn, Howard. *A People's History of the United States*. New York: Harper Perennial Modern Classics, 2005. Chapter 13, *"The Socialist Challange"* pp. 321-357 ISBN 0060838655

[84] Zinn, Howard. *A People's History of the United States*. New York: Harper Perennial Modern Classics, 2005. p. 327 ISBN 0060838655

[85] Foner, Eric; Garraty, John A. (1991). *The Reader's Companion to American History*. New York: Houghton Mifflin. p. 576. ISBN 0-395-51372-3.

[86] McDuffie, Jerome; Piggrem, Gary Wayne; Woodworth, Steven E. (2005). *U.S. History Super Review*. Piscataway, NJ: Research & Education Association. p. 418. ISBN 0-7386-0070-9.

[89] See, for example, the aircraft production figures cited in : in 1940, the Americans produced 6,019 aircraft; in 1941, 19,433; in 1942, almost 48,000; in 1943, an astonishing 85,898. Even more were to come in 1944. For comparison, in 1941, the German war economy could only manage about 12,000 aircraft.

[90] p. 2.

[91] Kennedy, Paul (1989). *The Rise and Fall of the Great Powers*. New York: Vintage. p. 358. ISBN 0-679-72019-7.
Indeed, World War II ushered in the zenith of US power in what came to be called the American Century, as , indicates: "Truman presided over the greatest military and economic power the world had ever known. War production had lifted the United States out of the Great Depression and had inaugurated an era of unimagined prosperity. Gross national product increased by 60 percent during the war, total earnings by 50 percent. Despite social unrest, labor agitation, racial conflict, and teenage vandalism, Americans had more discretionary income than ever before. Simultaneously, the US government had built up the greatest war machine in human history. By the end of 1942, the United States was producing more arms than all the Axis states combined, and, in 1943, it made almost three times more armaments than did the Soviet Union. In 1945, the United States had two-thirds of the world's gold reserves, three-fourths of its invested capital, half of its shipping vessels, and half of its manufacturing capacity. Its GNP was three times that of the Soviet Union and more than five times that of Britain. It was also nearing completion of the atomic bomb, a technological and production feat of huge costs and proportions."

[93] Pacific War Research Society (2006). *Japan's Longest Day*. New York: Oxford University Press. ISBN 4-7700-2887-3.

[97]

[98] Social Security History (http://www.ssa.gov/history/lbjsm.html), the United States Social Security Administration

[99] Soss, 2010, p.277

[100] Fraser, 1989

[101] Ferguson, 1986, pp.43-53

[102] Williams, pp.325–331

[103]

[104]

[105] Hayes, 2009

[106] US History.org, 2013

[107] Burns, 2008

[108] Levy, 1996, p.442

[109] Wallander, 2003

[110]

[112]

[113]

[114]

[115]

[118]

[132] Daynes & Sussman, 2010, pp.3, 72, 74-76, 78

[133] Hays, Samuel P. (2000). *A History of Environmental Politics since 1945.*

[134] Rothman, Hal K. (1998).*The Greening of a Nation? Environmentalism in the United States since 1945*

[135] Turner, James Morton (2012). *The Promise of Wilderness*

[136] http://www.census.gov/prod/cen2010/briefs/c2010br-02.pdf

[140] "U.S. Legal Permanent Residents: 2012" (http://www.dhs.gov/sites/default/files/publications/ois_lpr_fr_2012_2.pdf). Office of Immigration Statistics *Annual Flow Report.*

[144] National Survey of Sexual Health and Behavior (http://www.nationalsexstudy.indiana.edu/). Retrieved January 6, 2013.

[146] U.S. Census Bureau: "U.S. Census Bureau Delivers Final State 2010 Census Population Totals for Legislative Redistricting" (http://www.census.gov/2010census/news/releases/operations/cb11-cn123.html) see custom table, 2nd worksheet

[153] "Language Spoken at Home by the U.S. Population, 2010", American Community Survey, U.S. Census Bureau, in *World Almanac and Book of Facts 2012*, p. 615.

[159] Media, Minorities, and Meaning: A Critical Introduction — Page 88, Debra L. Merskin – 2010

[168] Scheb, John M., and John M. Scheb II (2002). *An Introduction to the American Legal System*. Florence, KY: Delmar, p. 6. ISBN 0-7668-2759-3.

[177] Feldstein, Fabozzi, 2011, p.9

[178] Schultz, 2009, pp.164, 453, 503

[179] Schultz, 2009, p.38

[181] See 8 U.S.C. § 1101 (http://www.law.cornell.edu/uscode/8/1101.html)(a)(36) and 8 U.S.C. (a)(38) U.S. Federal Code, Immigration and Nationality Act. 8 U.S.C. § 1101a (http://www.law.cornell.edu/uscode/8/1101a.html)

[183] US General Accounting Office, U.S. Insular Areas. Application of the U.S. Constitution (http://www.gao.gov/assets/230/224900.pdf). November 1997. p. 9. Appendix I, p. 23-38. Retrieved April 29, 2013.

[231] For the latest data, see For other estimates of the incarceration rate in China and North Korea see

[239] Change is based on chained 2005 dollars. Quarterly growth is expressed as an annualized rate.

[246] Wright, Gavin, and Jesse Czelusta, "Resource-Based Growth Past and Present", in *Natural Resources: Neither Curse Nor Destiny*, ed. Daniel Lederman and William Maloney (World Bank, 2007), p. 185. ISBN 0-8213-6545-2.

[259] "Personal Consumption Expenditures (PCE)/Gross Domestic Product (GDP)" (http://research.stlouisfed.org/fred2/graph/?g=hh3) *FRED Graph*, Federal Reserve Bank of St. Louis

[263] "Household Income for States: 2010 and 2011" (http://www.census.gov/prod/2012pubs/acsbr11-02.pdf) *United States Census, American Community Survey Briefs*, September 2012, Appendix Table 1, page 5

[269] Yen, Hope (28 July 2013). 80 Percent Of U.S. Adults Face Near-Poverty, Unemployment: Survey (http://www.huffingtonpost.com/2013/07/28/poverty-unemployment-rates_n_3666594.html). *The Huffington Post.* Retrieved 28 July 2013.

[271] "Extreme Poverty in the United States, 1996 to 2011" (http://npc.umich.edu/publications/policy_briefs/brief28/policybrief28.pdf) *National Poverty Center*, February 2012

[274] Gould, Elise and Wething, Hilary (24 July 2012). "U.S. poverty rates higher, safety net weaker than in peer countries." (http://www.epi.org/publication/ib339-us-poverty-higher-safety-net-weaker/) *Economic Policy Institute.* Retrieved 15 July 2013.

[276] Orr, D. (November–December 2004). "Social Security Isn't Broken: So Why the Rush to 'Fix' It?" In C. Sturr and R. Vasudevan, eds. (2007). *Current Economic Issues*. Boston: Economic Affairs Bureau.

[284] THE 2012 LEGATUM PROSPERITY INDEX TABLE RANKINGS (http://www.prosperity.com/Ranking.aspx). Legatum Institute, 2012.

[287] " Americans' wealth drops $1.3 trillion (http://money.cnn.com/2009/06/11/news/economy/Americans_wealth_drops/?postversion=2009061113)". CNN. June 11, 2009.

[288] " U.S. household wealth falls $11.2 trillion in 2008 (http://www.reuters.com/article/idUSTRE52B58720090312)". Reuters. March 12, 2009.

[324] For more detail on U.S. literacy, see A First Look at the Literacy of America's Adults in the 21st century (http://nces.ed.gov/NAAL/PDF/2006470.PDF), U.S. Department of Education (2003).

[328] "Slideshow: Most obese countries". (http://www.reuters.com/news/pictures/slideshow?articleId=USRTXT3DK#a=1) *Reuters*. Retrieved November 22, 2012.

[335] *OECD Health Data 2000: A Comparative Analysis of 29 Countries* [CD-ROM] (OECD: Paris, 2000). See also

[339] "Out-of-Pocket Spending in the Last Five Years of Life" (http://link.springer.com/article/10.1007/s11606-012-2199-x) *Journal of General Internal Medicine*, February 2013, Volume 28, Issue 2, pp 304–309

[343] Thompson, William; Hickey, Joseph (2005). *Society in Focus*. Boston: Pearson. ISBN 0-205-41365-X.

[344] Fiorina, Morris P.; Peterson, Paul E. (2000). *The New American Democracy*. London: Longman, p. 97. ISBN 0-321-07058-5.

[345] Holloway, Joseph E. (2005). *Africanisms in American Culture*, 2d ed. Bloomington: Indiana University Press, pp. 18–38. ISBN 0-253-34479-4. Johnson, Fern L. (1999). *Speaking Culturally: Language Diversity in the United States*. Thousand Oaks, Calif., London, and New Delhi: Sage, p. 116. ISBN 0-8039-5912-5.

[348] Gould, Elise (10 October 2012). "U.S. lags behind peer countries in mobility." (http://www.epi.org/publication/usa-lags-peer-countries-mobility/) *Economic Policy Institute.* Retrieved 15 July 2013.

[349] CAP: *Understanding Mobility in America* (http://www.americanprogress.org/issues/2006/04/b1579981.html). April 26, 2006

[355] *Village Voice*: 100 Best Films of the 20th century (2001) (http://www.filmsite.org/villvoice.html). Filmsite.

[356] *Sight and Sound* Top Ten Poll 2002 (http://www.bfi.org.uk/sightandsound/topten/poll/critics-long.html).. British Film Institute. Retrieved June 19, 2007.

[361] Biddle, Julian (2001). *What Was Hot!: Five Decades of Pop Culture in America*. New York: Citadel, p. ix. ISBN 0-8065-2311-5.

[362] Bloom, Harold. 1999. *Emily Dickinson*. Broomall, PA: Chelsea House. p. 9. ISBN 0-7910-5106-4.

[364] Quinn, Edward (2006). *A Dictionary of Literary and Thematic Terms*. Infobase, p. 361. ISBN 0-8160-6243-9. Seed, David (2009). *A Companion to Twentieth-Century United States Fiction*. Chichester, West Sussex: John Wiley and Sons, p. 76. ISBN 1-4051-4691-5. Meyers, Jeffrey (1999). *Hemingway: A Biography*. New York: Da Capo, p. 139. ISBN 0-306-80890-0.
[367] Brown, Milton W. (1988 1963). *The Story of the Armory Show*. New York: Abbeville. ISBN 0-89659-795-4.
[369] Smith, 2004, pp.131-132
[370] Levenstein, 2003, pp.154–55
[371] Boslaugh, Sarah (2010). "Obesity Epidemic", in *Culture Wars: An Encyclopedia of Issues, Viewpoints, and Voices*, ed. Roger Chapman. Armonk, N.Y.: M. E. Sharpe, pp. 413–14. ISBN 978-0-7656-1761-3.
[373] Global sports market to hit $141 billion in 2012 (http://www.reuters.com/article/2008/06/18/us-pwcstudy-idUSN1738075220080618). Reuters. Retrieved on 2013-07-24.
[374] Maccambridge, Michael (2004). *America's Game: The Epic Story of How Pro Football Captured a Nation*. New York: Random House. ISBN 0-375-50454-0.
[378] Norway is first.

## Bibliography

• , Book (http://books.google.com/books/about/Making_America_Volume_1_To_1877_A_Histor.html?id=cyEI21RClZkC)
• a
• , Book (http://books.google.com/books?id=9KT3lI76-0cC&source=gbs_navlinks_s)
• , Book (http://books.google.com/books?id=FS3TnrLu7y8C&source=gbs_navlinks_s)
• , Book (http://books.google.com/books?id=edYbAZ7ECEoC&source=gbs_navlinks_s)
• , Book (http://books.google.com/books?id=Dphzw2cbaoQC&vq=1670&source=gbs_navlinks_s)
• , Book (http://books.google.com/books?id=HIOQ1FYHtcYC&source=gbs_navlinks_s)
• , Book (http://books.google.com/books?id=Juc4fb1Fx1cC&source=gbs_navlinks_s)
• , Book (http://books.google.com/books?id=mbV6kPGO4OAC&source=gbs_navlinks_s)
• , Book (http://books.google.com/books/about/An_Empire_of_Wealth.html?id=rmsUs_KDgHAC)
• , eBook version (http://books.google.com/books/about/A_History_of_the_American_People.html?id=RXSVQjz1_tMC)
• , Book (http://books.google.com/books?id=0q2GwESzhUwC&vq=slave&source=gbs_navlinks_s)
• , Book (http://books.google.com/books?id=laxSyAp89G4C&source=gbs_navlinks_s)
• , Book (http://books.google.com/books?id=gvIeAAAAMAAJ&source=gbs_navlinks_s)
• , PDF (http://cascourses.uoregon.edu/geog471/wp-content/uploads/2010/01/Lemon.pdf)
• , E'Book (http://books.google.com/books?id=UYpVAAAAYAAJ&source=gbs_navlinks_s)
• , Book (http://books.google.com/books?id=24mIQuLBuN8C&pg=PA44)
• , Book (http://www.umich.edu/~historyj/docs/2007-fall/Interracial_Marriage_in_Early_America_Mann.pdf)
• eBook version (http://books.google.com/books?id=_EFbS_7fFcYC&printsec=frontcover#v=onepage&q&f=false)
• , Book (http://books.google.com/books?id=qqxK4KlqKYMC&source=gbs_navlinks_s)
• , Book (http://books.google.com/books?id=iyZMeGMgIgEC&vq=triangle&source=gbs_navlinks_s)
• , Book (http://books.google.com/books/about/The_House.html?id=CAM6J6IoQFQC)
• , Book (http://books.google.com/books?id=vX-fYvoAeHwC&source=gbs_navlinks_s)
• , E'Book (http://books.google.com/books?id=G7AJAAAAIAAJ&source=gbs_navlinks_s)
• , Book (http://books.google.com/books?id=QlemwRTsY20C&vq=census&source=gbs_navlinks_s)
• , Book (http://books.google.com/books?id=f7m713xwK58C&vq=twenty+seven+times&source=gbs_navlinks_s)
• , Book (http://books.google.com/books?id=N6lrAmPlbvIC&pg=PA79&dq=#v=onepage&q&f=false)
• , Book (http://books.google.com/?id=JttyjBoyb3AC&lpg=PA12)
• , Article (http://www.jstor.org/discover/10.2307/2652029?uid=3739560&uid=2129&uid=2&uid=70&uid=4&uid=3739256&purchase-type=article&accessType=RR&sid=21102125115943&

showMyJstorPss=false&seq=1&showAccess=true)

- , Book (http://books.google.com/books?id=NPoAQRgkrOcC&vq=1670&source=gbs_navlinks_s)
- , Book (http://books.google.com/?id=9iQYSQ9y60MC&lpg=PA49)
- , Book (http://books.google.com/books/about/History_of_the_American_Economy_With_Acc.html?id=lyhI1q_E4G0C)

### Website sources

- , Book (http://books.google.com/books?id=2MjxPJ9W4gwC&source=gbs_navlinks_s)

## External links

- United States (https://www.cia.gov/library/publications/the-world-factbook/geos/us.html) entry at *The World Factbook*
- United States (http://www.bbc.co.uk/news/world-us-canada-16761057), from the BBC News
- Key Development Forecasts for the United States (http://www.ifs.du.edu/ifs/frm_CountryProfile.aspx?Country=US) from International Futures

Government

- Official U.S. Government Web Portal (http://www.usa.gov/) Gateway to government sites
- House (http://www.house.gov/) Official site of the United States House of Representatives
- Senate (http://www.senate.gov/) Official site of the United States Senate
- White House (http://www.whitehouse.gov/) Official site of the President of the United States
- Supreme Court (http://www.supremecourt.gov/) Official site of the Supreme Court of the United States

History

- Historical Documents (http://www.nationalcenter.org/HistoricalDocuments.html) Collected by the National Center for Public Policy Research
- U.S. National Mottos: History and Constitutionality (http://www.religioustolerance.org/nat_mott.htm) Analysis by the Ontario Consultants on Religious Tolerance
- USA (http://www.historicalstatistics.org/index2.html) Collected links to historical data

Maps

- National Atlas of the United States (http://nationalatlas.gov/) Official maps from the U.S. Department of the Interior
- Wikimedia Atlas of the United States
- Measure of America (http://www.measureofamerica.org/maps/) A variety of mapped information relating to health, education, income, and demographics for the U.S.

# Pennsylvania

**Pennsylvania** i/ˌpɛnsɪlˈveɪnjə/, officially the **Commonwealth of Pennsylvania**, is a US state that is located in the Northeastern and Mid-Atlantic regions of the United States, and the Great Lakes region. The state borders Delaware to the southeast, Maryland to the south, West Virginia to the southwest, Ohio to the west, Lake Erie and Ontario, Canada to the northwest, New York to the north and New Jersey to the east. The Appalachian Mountains run through the middle of the state.

Pennsylvania is the 33rd most extensive, the 6th most populous, and the 9th most densely populated of the 50 United States. The state's four most populous cities are Philadelphia, Pittsburgh, Allentown, and Erie. The state capital is Harrisburg. Pennsylvania has 51 miles ( km) of coastline along Lake Erie and 57 miles ( km)[] of shoreline along the Delaware Estuary. The state is one of the 13 original founding states of the U.S.

## Geography

Further information: Geography of Pennsylvania and List of counties in Pennsylvania

Pennsylvania is 170 miles ( km) north to south and 283 miles ( km) east to west.[] Of a total 46055 square miles ( $km^2$), 44817 square miles ( $km^2$) are land, 490 square miles ( $km^2$) are inland waters, and 749 square miles ( $km^2$) are waters in Lake Erie.[1] It is the 33rd largest state in the United States.[2] Pennsylvania has 51 miles ( km)[] of coastline along Lake Erie and 57 miles ( km)[] of shoreline along the Delaware Estuary.

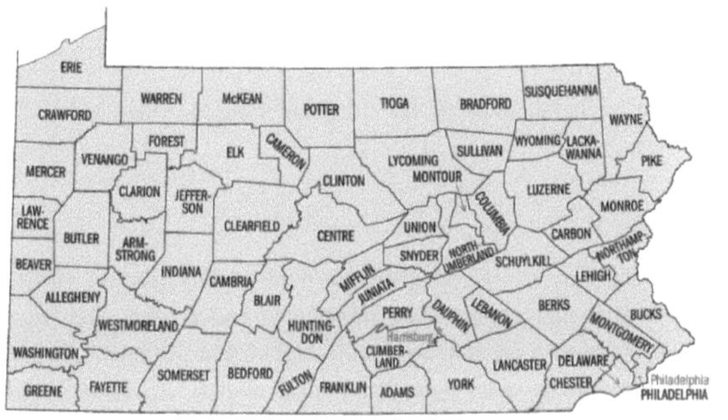

Counties of Pennsylvania.

The bounds of the state are the Mason–Dixon line (39° 43' N) to the south, the Twelve-Mile Circle on the Pennsylvania-Delaware border, the Delaware River to the east, 80° 31' W to the west and the 42° N to the north, with the exception of a short segment on the western end, where a triangle extends north to Lake Erie.

Pennsylvania borders six other states: New York to the north; New Jersey to the east; Delaware and Maryland to the southeast; West Virginia to the southwest, and Ohio to the west. Pennsylvania also shares a water border with the Canadian province of Ontario. Of the original Thirteen

Colonies, Pennsylvania is the only state that does not border the Atlantic Ocean.

It has the cities of Philadelphia, Reading, Lebanon and Lancaster in the southeast, Pittsburgh in the southwest, the tri-cities of Allentown, Bethlehem, and Easton in the central east (known as the Lehigh Valley), the tri-cities of Scranton, Wilkes-Barre, and Hazleton in the northeast, and Erie in the northwest. Williamsport serves as the "hub" of the commonwealth's north-central region, with York and the state capital Harrisburg on the Susquehanna River in the central region of the commonwealth.

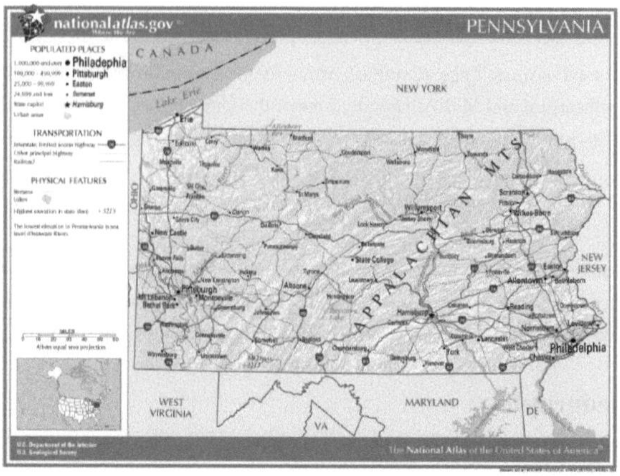

Map of Pennsylvania, showing major cities and roads.

There are 5 regions. They are the Allegheny Plateau, Ridge and Valley, Atlantic Coastal Plain, Piedmont, and the Erie Plain.

## Climate

Pennsylvania's diverse topography also produces a variety of climates, though the entire state experiences cold winters and humid summers. Straddling two major zones, the majority of the state, with the exception of the southeastern corner, has a humid continental climate (Köppen climate classification *Dfa*). Greater Philadelphia has some characteristics of the humid subtropical climate (Köppen *Cfa*) that covers much of Delaware and Maryland to the south.

Moving toward the mountainous interior of the state, the winter climate becomes markedly colder, the number of cloudy days increases, and snowfall amounts are greater. Western areas of the state, particularly locations near Lake Erie, can receive over 100 inches ( cm) of snowfall annually, and the entire state receives plentiful precipitation throughout the year. The state may be subject to severe weather from spring through summer into fall; an average of 10 tornadoes touch down each year in the state.

Monthly Normal High and Low Temperatures For Various Pennsylvania Cities in °F

| City | Jan. | Feb. | Mar. | Apr. | May. | Jun. | Jul. | Aug. | Sep. | Oct. | Nov. | Dec. |
|---|---|---|---|---|---|---|---|---|---|---|---|---|
| **Scranton** | 33/19 | 37/21 | 46/28 | 59/38 | 70/48 | 78/56 | 82/61 | 80/60 | 72/52 | 61/41 | 49/33 | 38/24 |
| **Erie** | 34/21 | 36/21 | 44/27 | 56/38 | 67/48 | 76/58 | 80/63 | 79/62 | 72/56 | 61/45 | 50/37 | 38/27 |
| **Pittsburgh** | 36/21 | 39/23 | 49/30 | 62/40 | 71/49 | 79/58 | 83/63 | 81/62 | 74/54 | 63/43 | 51/35 | 39/25 |

| | | | | | | | | | | | | |
|---|---|---|---|---|---|---|---|---|---|---|---|---|
| **Harrisburg** | 37/23 | 41/25 | 50/33 | 62/42 | 72/52 | 81/62 | 85/66 | 83/64 | 76/56 | 64/45 | 53/35 | 41/27 |
| **Philadelphia** | 40/26 | 44/28 | 53/34 | 64/44 | 74/54 | 83/64 | 87/69 | 85/68 | 78/60 | 67/48 | 56/39 | 45/30 |
| **Allentown** | 36/20 | 40/22 | 49/29 | 61/39 | 72/48 | 80/58 | 84/63 | 82/61 | 75/53 | 64/41 | 52/33 | 40/24 |
| [3], [4], [5], [6], [7], | | | | | | | | | | | | |

## History

Before the Commonwealth was settled by Europeans, the area was home to the Delaware (also known as Lenni Lenape), Susquehannock, Iroquois, Eriez, Shawnee, and other American Indian Nations.[8] Both the Dutch and the English claimed both sides of the Delaware River as part of their colonial lands in America.[9] [10] [11] The Dutch were the first to take possession, which has impact on the history of Pennsylvania.[12]

By June 3, 1631, the Dutch had started the settlement of the Delmarva Peninsula by establishing the Zwaanendael Colony on the site of present day Lewes, Delaware.[13] In 1638, Sweden heated up the issue by establishing the New Sweden Colony, centered on Fort Christina, on the site of present day Wilmington, Delaware. New Sweden claimed and, for the most part, controlled the lower Delaware River region (Parts of present Delaware, New Jersey and Pennsylvania) but settled few colonists there.[14] [15]

On March 12, 1664, King Charles II of England gave James, Duke of York a grant that included all of the lands included in the original Virginia Company of Plymouth Grant as well as other lands. This grant was – again – in conflict with the Dutch claim for New Netherland, which included parts of today's Pennsylvania.[16]

*Penn's Treaty with the Indians*, by Edward Hicks.

On June 24, 1664, The Duke of York sold the portion of his large grant that included present day New Jersey to John Berkeley and George Carteret for a proprietary colony. The land was not yet in British possession, but the sale boxed in the portion of New Netherland on the West side of the Delaware River. The British conquest of New Netherland was commenced on August 29, 1664, when New Amsterdam was coerced to surrender, facing the cannons on British ships in New York Harbor.[17] [18] This conquest continued, and was completed in October 1664, when the British captured Fort Casimir in what today is New Castle, Delaware.

The Peace of Breda between England, France and the Netherlands confirmed the British conquest on July 21, 1667,[19] [20] although there were temporary reversions.

On September 12, 1672, as part of the Third Anglo—Dutch War, the Dutch re-conquered New York Colony/New Amsterdam, the Dutch established three County Courts which went on to become original Counties in present day Delaware and Pennsylvania. The one that later transferred to Pennsylvania was Upland.[21] This was partially reversed on February 9, 1674, when the Treaty of Westminster ended the Third Anglo-Dutch War, and reverted all political situations to the *status quo ante bellum*. The British retained the Dutch Counties with their Dutch names.[22] By June 11, 1674, New York reasserted control over the outlying colonies, including Upland, but the names started to be changed to British names by November 11, 1674.[23] Upland was partitioned on November 12, 1674, producing the general outline of the current border between Pennsylvania and Delaware.[24]

John Dickinson

On February 28, 1681, Charles II granted a land charter[25] to William Penn to repay a debt of £16,000[26] (around £2,100,000 in 2008, adjusting for retail inflation)[27] owed to William's father, Admiral William Penn. This was one of the largest land grants to an individual in history.[] It was called Pennsylvania. William Penn, who wanted it called New Wales or Sylvania, was embarrassed at the change, fearing that people would think he had named it after himself, but King Charles would not rename the grant.[28] Penn established a government with two innovations that were much copied in the New World: the county commission and freedom of religious conviction.[]

What had been Upland on what became the Pennsylvania side of the Pennsylvania-Delaware Border was renamed as Chester County when Pennsylvania instituted their colonial governments on March 4, 1681.[29] [30] The Quaker leader William Penn had signed a peace treaty with Tammany, leader of the Delaware tribe, beginning a long period of friendly relations between the Quakers and the Indians.[31] Additional treaties between Quakers and other tribes followed. The treaty of William Penn was never violated.[32]

## 18th century

Between 1730 and when it was shut down by Parliament with the Currency Act of 1764, the Pennsylvania Colony made its own paper money to account for the shortage of actual gold and silver. The paper money was called Colonial Scrip. The Colony issued "bills of credit", which were as good as gold or silver coins because of their legal tender status. Since they were issued by the government and not a banking institution, it was an interest-free proposition, largely defraying the expense of the government and therefore taxation of the people. It also promoted general employment and prosperity, since the Government used discretion and did not issue too much to inflate the currency. Benjamin Franklin had a hand in creating this currency, of which he said its utility was never to be disputed, and it also met with the "cautious approval" of Adam Smith.[33]

After the Stamp Act Congress of 1765, Delegate John Dickinson of Philadelphia, Pennsylvania, wrote the *Declaration of Rights and Grievances.* The Congress was the first meeting of the thirteen colonies, called at the request of the Massachusetts Assembly, but only nine colonies sent delegates.[34] Dickinson then wrote *Letters from a Farmer in Pennsylvania, To the Inhabitants of the British Colonies*, which were published in the Pennsylvania Chronicle between December 2, 1767, and February 15, 1768.[35]

President's House, Philadelphia. The Masters-Penn mansion housed Pennsylvania's governor in the early 1770s. It later served as the presidential mansion of George Washington and John Adams, 1790–1800, while Philadelphia was the temporary national capital.

When the Founding Fathers of the United States convened in Philadelphia in 1774, 12 colonies sent representatives to the First Continental Congress.[36] The Second Continental Congress, which also met in Philadelphia (in May, 1775), drew up and signed the Declaration of Independence in Philadelphia,[37] but when that city was captured by the British, the Continental Congress escaped westward, meeting at the Lancaster courthouse on Saturday, September 27, 1777, and then to York. There they drew up the Articles of Confederation that formed 13 independent colonies into a new nation. Later, the Constitution was written, and Philadelphia was once again chosen to be cradle to the new American Nation.[38]

Pennsylvania became the second state to ratify the U.S. Constitution on December 12, 1787,[39] five days after Delaware became the first.

Dickinson College of Carlisle was the first college founded in the United States. Established in 1773, the college was ratified five days after the Treaty of Paris on September 9, 1783. The school was founded by Benjamin Rush and named after John Dickinson.

For half a century, the Commonwealth's General Assembly (legislature) met at various places in the general Philadelphia area before starting to meet regularly in Independence Hall in Philadelphia for 63 years.[] But it needed a more central location, as for example the Paxton Boys massacres of 1763 had made the legislature aware. So, in 1799 the General Assembly moved to the Lancaster Courthouse,[] and finally in 1812 to Harrisburg.[]

The "Hills Capitol", used from 1821 until it burned down in 1897.

The General Assembly met in the old Dauphin County Court House until December 1821,[] when the Federal-style "Hills Capitol" (named for its builder, Stephen Hills, a Lancaster architect) was constructed on a hilltop land grant of four acres set aside for a seat of state government by the prescient, entrepreneurial son and namesake of John Harris, Sr., a Yorkshire native who had founded a trading post in 1705 and ferry (1733) on the east shore of the Susquehanna River.[40] The Hills Capitol burned down on February 2, 1897, during a heavy snowstorm, presumably because of a faulty flue.[] The General Assembly met at Grace Methodist Church on State Street (still standing) until the a new capitol could be built. Following an architectural selection contest that many alleged had been "rigged," Chicago architect Henry Ives Cobb was charged with designing and building a replacement building; however, the legislature had little money to allocate to the project, and a roughly finished, somewhat industrial building (the Cobb Capitol) was completed. The General Assembly refused to occupy the building. Political and popular indignation in 1901 prompted a second contest that was restricted to Pennsylvania architects, and Joseph Miller Huston of Philadelphia was chosen to design the present Pennsylvania State Capitol that incorporated Cobb's building into magnificent public work finished and dedicated in 1907.[]

The new state Capitol drew rave reviews.[] Its dome was inspired by the domes of St. Peter's Basilica in Rome and the United States Capitol.[] President Theodore Roosevelt called it "the most beautiful state Capital in the nation" and said, "It's the handsomest building I ever saw" at the dedication. In 1989, the *New York Times* praised it as "grand, even awesome at moments, but it is also a working building, accessible to citizens ... a building that connects with the reality of daily life".[]

Roosevelt's FERA camp for unemployed women, 1934

Pennsylvania accounts for nine percent of wooded areas in the United States. In 1923 President Calvin Coolidge established the Allegheny National Forest under the authority of the Weeks Act of 1911 in the northwest part of the state in Elk, Forest, McKean, and Warren Counties for the purposes of timber production and watershed protection in the Allegheny River basin. The Allegheny is the state's only national forest.

James Buchanan, of Franklin County, the only bachelor President of the United States,[] was the only one to be born in Pennsylvania. The Battle of Gettysburg—the major turning point of the Civil War—took place near Gettysburg.[41] An estimated 350,000 Pennsylvanians served in the Union Army forces along with 8,600 African American military volunteers.

Pennsylvania was also the home of the first commercially drilled oil well. In 1859, near Titusville, Pennsylvania, Edwin L. Drake successfully drilled the well, which led to the first major oil boom in United States history.

## Demographics

Further information: List of people from Pennsylvania

The United States Census Bureau estimates that the population of Pennsylvania was 12,763,536 on July 1, 2012, a 0.5% increase since the 2010 United States Census.[]

Of the people residing in Pennsylvania, 74.5% were born in Pennsylvania, 18.4% were born in a different US state, 1.5% were born Puerto Rico, U.S. Island areas, or born abroad to American parent(s), and 5.6% were foreign born.[42]

According to the 2010 Census, 81.9% of the population was White (79.2% non-Hispanic white), 11.3% was Black or African American, 0.3% American Indian and Alaska Native, 2.9% Asian, 1.9% from two or more races. 5.9% of the total population was of Hispanic or Latino origin (they may be of any race).[43]

As of 2011, 32.1% of Pennsylvania's population younger than age 1 were minorities.[44]

Pennsylvania's Hispanic population grew by 82.6% between 2000 and 2010, making it one of the largest increases in a state's Hispanic population. The significant growth of the Hispanic population is due to immigration to the state mainly from Puerto Rico, which is a US territory, but to a lesser extent from countries such as the Dominican Republic, Mexico, and various Central and South American nations, as well as from the wave of Hispanics leaving New York and New Jersey for safer and more affordable living. The Asian population swelled by almost 60%, which was fueled by Indian, Vietnamese, and Chinese immigration, as well the many Asian transplants moving to Philadelphia from New York. The rapid growth of this community has given Pennsylvania one of the largest Asian populations in the nation by numerical values. The Black and African American population grew by 13%, which was the largest increase in that population amongst the state's peers (New York, New Jersey, Ohio, Illinois, and Michigan).The White population declined by 0.7%, a trend that is beginning to reverse itself. Twelve other states saw decreases in their White populations.[45]

The center of population of Pennsylvania is located in Perry County, in the borough of Duncannon.[46]

As of 2006, Pennsylvania has an estimated population of 12,440,621, which is an increase of 35,273 from the previous year and an increase of 159,567 since the year 2000. Net migration to other states resulted in a decrease of 27,718, and immigration from other countries resulted in an increase of 127,007. Net migration to the Commonwealth was 98,289. Migration of native Pennsylvanians resulted in a decrease of 100,000 people. In 2006, 5.00% of Pennsylvanians were foreign born (621,480 people). The state has an estimated 2005 poverty rate of 11.9%.[] The state also has the 3rd highest proportion of elderly (65+) citizens in 2005.[]

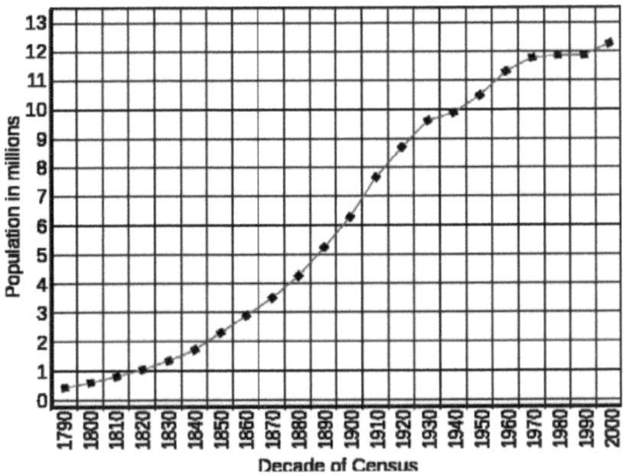


State population from 1790 to 2000

Foreign born Pennsylvanians are largely from Asia (36.0%), Europe (35.9%), Latin America (30.6%), Africa (5%), North America (3.1%), and Oceania (0.4%).

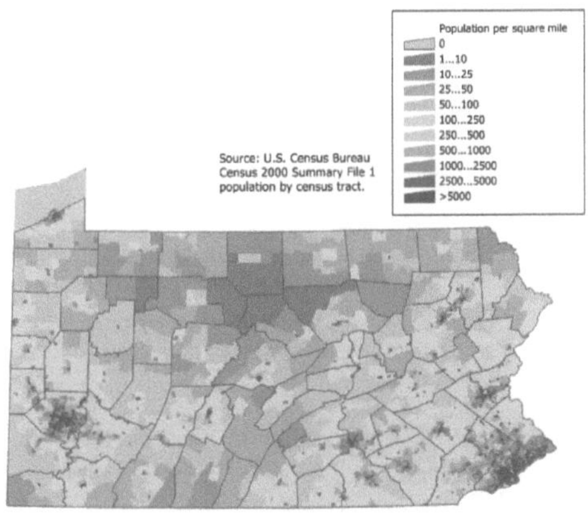


Pennsylvania's population distribution.

Pennsylvania's reported population of Hispanics, especially among the Asian, Hawaiian and White races, has markedly increased in recent years.[] The Hispanic population is greatest in Allentown, Lancaster, Reading, Hazleton, and around Philadelphia, with over 20% being Hispanic. It is not clear how much of this change reflects a changing population and how much reflects increased willingness to self-identify minority status. As of 2010, it is estimated that about 85% of all Hispanics in Pennsylvania live within a 150 miles ( km) radius of Philadelphia, with about 20% living within the city itself.

Pennsylvania's population was reported as 5.9% under 5 and 23.8% under 18, with 15.6% aged 65 or older. Females made up 51.7% of the population.[] The largest ancestry groups are listed below, expressed as a percentage of total people who responded with a particular ancestry for the 2006–2008 census:[47]

- 28.5% German
- 18.2% Irish

- 12.8% Italian
- 9.6% African
- 8.5% English
- 7.2% Polish
- 4.2% French Canadian
- 2.9% Puerto Rican[]
- 2.2% Dutch
- 2.0% Slovak
- 2.0% Scotch Irish
- 1.7% Scottish
- 1.6% Russian
- 1.5% Welsh
- 1.2% Hungarian
- 1.0% West Indian
- 1.0% Ukrainian
- 1.0% Mexican

The five largest estimated ancestry groups in Pennsylvania are: German (28.5%), Irish (18.2%), Italian (12.8%), English (8.5%) and Polish (7.2%).

## Demographics of Pennsylvania (csv) [48]

| By race | White | Black | AIAN* | Asian | NHPI* |
|---|---|---|---|---|---|
| **2000 (total population)** | 87.60% | 10.71% | 0.43% | 2.04% | 0.07% |
| **2000 (Hispanic only)** | 2.74% | 0.44% | 0.06% | 0.03% | 0.02% |
| **2005 (total population)** | 86.83% | 11.20% | 0.45% | 2.46% | 0.09% |
| **2005 (Hispanic only)** | 3.52% | 0.53% | 0.07% | 0.05% | 0.02% |
| **Growth 2000–05 (total population)** | 0.32% | 5.83% | 5.64% | 22.23% | 18.99% |
| **Growth 2000–05 (non-Hispanic only)** | -0.64% | 5.21% | 2.77% | 21.86% | 14.13% |
| **Growth 2000–05 (Hispanic only)** | 29.86% | 20.24% | 23.61% | 45.64% | 35.44% |

* AIAN is American Indian or Alaskan Native; NHPI is Native Hawaiian or Pacific Islander

| Historical populations | | |
|---|---|---|
| **Census** | **Pop.** | **%±** |
| **1790** | 434373 | — |
| **1800** | 602365 | 38.7% |
| **1810** | 810091 | 34.5% |
| **1820** | 1049458 | 29.5% |
| **1830** | 1348233 | 28.5% |
| **1840** | 1724033 | 27.9% |
| **1850** | 2311786 | 34.1% |
| **1860** | 2906215 | 25.7% |
| **1870** | 3521951 | 21.2% |
| **1880** | 4282891 | 21.6% |

| | | |
|---|---|---|
| **1890** | 5258113 | 22.8% |
| **1900** | 6302115 | 19.9% |
| **1910** | 7665111 | 21.6% |
| **1920** | 8720017 | 13.8% |
| **1930** | 9631350 | 10.5% |
| **1940** | 9900180 | 2.8% |
| **1950** | 10498012 | 6.0% |
| **1960** | 11319366 | 7.8% |
| **1970** | 11793909 | 4.2% |
| **1980** | 11863895 | 0.6% |
| **1990** | 11881643 | 0.1% |
| **2000** | 12281054 | 3.4% |
| **2010** | 12702379 | 3.4% |
| **Est. 2012** | 12763536 | 0.5% |
| Source: 1910–2010[49] | | |

## Languages

### Top 10 Non-English Languages Spoken in Pennsylvania

| Language | Percentage of population (as of 2010)[] |
|---|---|
| Spanish | 4.09% |
| Chinese (including Mandarin) and German (tied) | 0.47% |
| Italian | 0.43% |
| Pennsylvania Dutch | 0.40% |
| French | 0.34% |
| Russian and Vietnamese (tied) | 0.29% |
| Korean | 0.25% |
| Polish | 0.21% |
| Arabic | 0.20% |
| Hindi | 0.17% |

As of 2010, 90.15% (10,710,239) of Pennsylvania residents age 5 and older spoke English at home as a primary language, while 4.09% (486,058) spoke Spanish, 0.47% (56,052) Chinese (which includes Mandarin,) and German was spoken as a main language by 0.47% (55,891) of the population over the age of five. In total, 9.85% (1,170,628) of California's population age 5 and older spoke a mother language other than English.[]

### Pennsylvania Dutch

The term "Dutch," when referring to the language spoken by the Pennsylvania Dutch people, (Pennsylvania Dutch), means "German" or "Teutonic" rather than "Netherlander." Germans, in their own language, call themselves "Deutsch," which in English became, misleadingly, "Dutch."[50] The Pennsylvania Dutch language is a descendant of German, in the West Central German dialect family. Although it is still spoken as a first language among some Old Order Amish and Mennonites (principally in the Lancaster County area), the language is almost extinct as an everyday language among the non-religious, though a few words have passed into English usage.

## Religion

> The new sovereign also enacted several wise and wholesome laws for his colony, which have remained invariably the same to this day. The chief is, to ill–treat no person on account of religion, and to consider as brethren all those who believe in one God.
> – Voltaire, speaking of William Penn[51]

Of all the colonies, only Rhode Island had religious freedom as secure as in Pennsylvania, and one result was a wide religious diversity, which continues to this day.[52]

Pennsylvania's population in 2000 was 12,281,054. Of these, 8,448,193 were estimated to belong to some sort of organized religion. According to the Association of religion data archives at Pennsylvania State University, reliable data exists for 7,116,348 religious adherents in Pennsylvania in 2000 following 115 different faiths.[] While Pennsylvania has a very large Amish population, Holmes County, Ohio has the largest Amish population in the world.[53] While Pennsylvania owes its existence to Quakers and many of the older trappings of the Commonwealth are rooted in the teachings of the Religious Society of Friends (as they are officially known), practicing Quakers are a small minority today.

The religious affiliations of the people of Pennsylvania:[]

- Christianity – 80%
    - Protestant – 51%
        - Mainline Protestant – 25%
        - Evangelical Protestant – 18%
        - Black Protestant – 7%
        - Jehovah Witness – 1%
    - Roman Catholic – 29%
    - Orthodox – 1%
- Non-religious/ Unaffiliated – 13%
- Judaism – 2%
- Other religions – 1%
- Don't know/ refused – 1%

According to a 2012 Gallup poll, 40% of Pennsylvanians are very religious, 28% are moderately religious, and 32% are non-religious.[54]

## Economy

Bethlehem Steel's closed manufacturing facility in Bethlehem, Pennsylvania. This site later became a new multi-million dollar Sands Casino Resort in 2009 and became the site of the ArtsQuest Center at Steelstacks, Levitt Pavilion Steelstacks, and PBS 39 in 2011.

Pennsylvania's 2010 total gross state product (GSP) of $570 billion ranks the state 6th in the nation.[] If Pennsylvania were an independent country, its economy would rank as the 18th largest in the world. On a per-capita basis, Pennsylvania's per-capita GSP of $39,830 ranks 29th among the 50 states.[]

Philadelphia in the southeast corner, Pittsburgh in the southwest corner, Erie in the northwest corner, Scranton-Wilkes-Barre in the northeast corner, and Allentown-Bethlehem-Easton in the east central region are urban manufacturing centers. Much of the Commonwealth is rural; this dichotomy affects state politics as well as the state economy.[55] Philadelphia is home to six Fortune 500 companies,[] with more located in suburbs like King of Prussia; it's a leader in the financial[56] and insurance industry.

Pittsburgh is home to eight Fortune 500 companies, including U.S. Steel, PPG Industries, and H.J. Heinz.[] In all, Pennsylvania is home to fifty Fortune 500 companies.[] Erie is also home to GE Transportation Systems, which is the largest producer of train locomotives in the United States.

As in the US as a whole and in most states, the largest private employer in the Commonwealth is Wal-Mart, followed by the University of Pennsylvania.[57] [58]

As of April 2012, the state's unemployment rate is 7.4%.[59]

### Banking

The famous Crayola crayon company is based in Easton.

The first nationally chartered bank in the United States, the Bank of North America, was founded in 1781 in Philadelphia. After a series of mergers, the Bank of North America is part of Wells Fargo, which uses national charter 1.

Pennsylvania is also the home to the first nationally chartered bank under the 1863 National Banking Act. That year, the Pittsburgh Savings & Trust Company received a national charter and renamed itself the First National Bank of Pittsburgh as part of the National Banking Act. That bank is still in existence today as PNC Financial Services, and remains based in Pittsburgh. PNC is the state's largest bank, and the sixth-largest in the United States.

### Agriculture

Pennsylvania ranks 19th overall in agricultural production,[60] but 1st in mushrooms, 2nd in apples, 3rd in Christmas trees and layer chickens, 4th in nursery and sod, milk, corn for silage, grapes grown (including juice grapes),[] and horses production. It also ranks 8th in the nation in Winemaking.[]

### Gambling

Casino gambling was legalized in Pennsylvania in 2004. Currently, there are nine casinos across the state with three under construction or in planning. Only horse racing, slot machines and electronic table games were legal in Pennsylvania, although a bill to legalize table games was being negotiated in the fall of 2009.[61] Tables games such as poker, roulette, black jack and dice were finally approved by the state legislature in January 2010, being signed into law by the Governor on January 7. Sports betting is illegal.

Governor Ed Rendell had considered legalizing video poker machines in bars and private clubs in 2009, since an estimated 17,000 operate illegally across the state.[62] Under this plan, any establishment with a liquor license would be allowed up to 5 machines. All machines would be connected to the state's computer system, like commercial casinos. The state would impose a 50% tax on net gambling revenues, after winning players have been paid, with the remaining 50% going to the establishment owners.

### Film

The Pennsylvania Film Production Tax Credit began in 2004 and stimulated the development of a film industry in the state.

## Governance

**Presidential elections results[63]**

| Year | Republican | Democratic |
|---|---|---|
| 2012 | 46.58% *2,680,434* | **51.97%** *2,990,274* |
| 2008 | 44.15% *2,655,885* | **54.47%** *3,276,363* |
| 2004 | 48.42% *2,793,847* | **50.92%** *2,938,095* |
| 2000 | 46.43% *2,281,127* | **50.60%** *2,485,967* |
| 1996 | 39.97% *1,801,169* | **49.17%** *2,215,819* |
| 1992 | 36.13% *1,791,841* | **45.15%** *2,239,164* |
| 1988 | **50.70%** *2,300,087* | 48.39% *2,194,944* |
| 1984 | **53.34%** *2,584,323* | 45.99% *2,228,131* |
| 1980 | **49.59%** *2,261,872* | 42.48% *1,937,540* |
| 1976 | 47.73% *2,205,604* | **50.40%** *2,328,677* |
| 1972 | **59.11%** *2,714,521* | 39.13% *1,796,951* |
| 1968 | 44.02% *2,090,017* | **47.59%** *2,259,405* |
| 1964 | 34.70% *1,673,657* | **64.92%** *3,130,954* |
| 1960 | 48.74% *2,439,956* | **51.06%** *2,556,282* |

| Voter Registration and Party Enrollment as of June 28, 2010[64] | | | |
|---|---|---|---|
| **Party** | | **Number of Voters** | **Percentage** |
| | Democratic | 4,309,604 | 51.00% |
| | Republican | 3,122,036 | 36.95% |
| | Unaffiliated | 492,077 | 5.82% |
| | Other Parties | 525,962 | 6.22% |
| **Total** | | **8,449,679** | **100%** |

## Government

Pennsylvania has had five constitutions during its statehood:[] 1776, 1790, 1838, 1874, and 1968 [65]. Prior to that, the province of Pennsylvania was governed for a century by a Frame of Government, of which there were four versions: 1682, 1683, 1696, and 1701.[] The capital of Pennsylvania is Harrisburg. The legislature meets in the State Capitol there.

Since 1992, Pennsylvania has been trending Democratic in Presidential elections (though the Pittsburgh metropolitan area trended more Republican in the 2008 Presidential election), voting for Bill Clinton twice by large margins, and slightly closer in 2000 for Al Gore. In the 2004 Presidential Election, Senator John F. Kerry beat President George W. Bush in Pennsylvania 2,938,095 (50.92%) to 2,793,847 (48.42%). Most recently, in the 2008 Presidential Election, Democrat Barack Obama defeated Republican John McCain in Pennsylvania, 3,184,778 (54%) to 2,584,088 (44%). The state holds 20 electoral votes.[]

In recent elections, Pennsylvania has leaned Democratic. The state has voted for the Democratic ticket for president in every election since 1992 and during the 2008 election campaign a recruitment drive saw registered Democrats outnumber registered Republicans by 1.2 million. However, Pennsylvania has a history of electing Republican senators. From 2009 to 2011, the state was represented by two Democratic senators for the first time since 1947. In 2010, Republicans recaptured a U.S. Senate seat as well as a majority of the state's congressional seats, control of both chambers of the state legislature and the governor's mansion. Democratic political consultant James Carville once pejoratively described Pennsylvania as "Philadelphia in the east, Pittsburgh in the west and Alabama in the middle".

### Governor

The current Governor is Tom Corbett, former Attorney General of Pennsylvania. The other elected officials composing the executive branch are the Lieutenant Governor Jim Cawley, Attorney General Kathleen Kane, Auditor General Eugene DePasquale, and State Treasurer Robert McCord.[66]

### General Assembly

Pennsylvania has a bicameral legislature set up by Commonwealth's constitution in 1790. The original Frame of Government of William Penn had a unicameral legislature.[67] The General Assembly includes 50 Senators[68] and 203 Representatives.[69] Joe Scarnati is currently President Pro Tempore of the State Senate,[70] Dominic Pileggi the Majority Leader,[71] and Jay Costa the Minority Leader.[72] Sam Smith is Speaker of the House of Representatives,[73] with Mike Turzai as Majority Leader[74] and Frank Dermody as Minority Leader.[75] As of the 2012 elections, the Republicans hold the majority in the State House and Senate.

### Judiciary

Pennsylvania is divided into 60 judicial districts,[] most of which (except Philadelphia) have magisterial district judges (formerly called district justices and justices of the peace), who preside mainly over preliminary hearings in felony and misdemeanor offenses, all minor (summary) criminal offenses, and small civil claims.[] Most criminal and civil cases originate in the Courts of Common Pleas, which also serve as appellate courts to the district judges and for local agency decisions.[] The Superior Court hears all appeals from the Courts of Common Pleas not expressly designated to the Commonwealth Court or Supreme Court. It also has original jurisdiction to review warrants for wiretap surveillance.[] The Commonwealth Court is limited to appeals from final orders of certain state agencies and certain designated cases from the Courts of Common Pleas.[] The Supreme Court of Pennsylvania is the final appellate court. All judges in Pennsylvania are elected; the chief justice is determined by seniority.[]

### Taxation

Pennsylvania has the 10th highest tax burden in the United States.[76] Residents pay a total of $83.7 billion in state and local taxes with a per capita average of $6,640 annually. Residents share 76% of the total tax burden. Many state politicians have tried to increase the share of taxes paid by out of state sources. Suggested revenue sources include taxing natural gas drilling as Pennsylvania is the only state without such a taxation on gas drilling.[77] Additional revenue prospects include trying to place tolls on interstate highways; specifically Interstate 80 which is used heavily by out of state commuters with high maintenance costs.[78]

Sales tax provides 39% of Commonwealth's revenue; personal income tax 34%; motor vehicle taxes about 12%, and taxes on cigarettes and alcohol beverage 5%.[79] Personal income tax is a flat 3.07%. An individual's taxable income is based on the following eight types of income: compensation (salary); interest; dividends; net profits from the operation of a business, profession or farm; net gains or income from the dispositions of property; net gains or income from rents, royalties, patents and copyrights; income derived through estates or trusts; and gambling and lottery winnings (other than Pennsylvania Lottery winnings).[80]

Counties, municipalities, and school districts levy taxes on real estate. In addition, some local bodies assess a wage tax on personal income. Generally, the total wage tax rate is capped at 1% of income but some municipalities with home rule charters may charge more than 1%. Thirty-two of the Commonwealth's sixty-seven counties levy a personal property tax on stocks, bonds, and similar holdings.

## Representation in the 113th Congress

Pennsylvania's two U.S. Senators in the 113th Congress are Bob Casey, Jr. and Pat Toomey.

Pennsylvania's U.S. Representatives for the term beginning January 2013 are Bob Brady (1st), Chaka Fattah (2nd), Mike Kelly (3rd), Scott Perry (politician) (4th), Glenn "G.T." Thompson (5th), Jim Gerlach (6th), Pat Meehan (7th), Mike Fitzpatrick (8th), Bill Shuster (9th), Tom Marino (10th), Lou Barletta (11th), Keith Rothfus (12th), Allyson Schwartz (13th), Mike Doyle (14th), Charlie Dent (15th), Joe Pitts (16th), Matt Cartwright (17th), Tim Murphy (18th).[81]

*See map of congressional districts*

## Municipalities

The skyline of Philadelphia, the largest city in Pennsylvania.

Pennsylvania is divided into 67 counties.[82] Counties are further subdivided into municipalities that are either incorporated as cities, boroughs, or townships.[83] One county, Philadelphia County, is coterminous with the city of Philadelphia after it was consolidated in 1854.

There are a total of 56 cities in Pennsylvania, which are classified, by population, as either first, second, or third class cities.[82] [84] Philadelphia, Pennsylvania's largest city, has a population of 1,547,297 and is the state's only first class city.[83] Pittsburgh (305,647) and Scranton (76,072) are second class and second class 'A' cities, respectively.[83]

The skyline of Pittsburgh, the second largest city in Pennsylvania.

The rest of the cities, like the third and fourth largest—Allentown (107,815) and Erie (103,571)—to the smallest—Parker with a population of only 738—are third class cities.[85] First and second class cities are governed by a "strong mayor" form of mayor–council government, whereas third class cities are governed by either a "weak mayor" form of government or a council–manager government.[83]

Boroughs are generally smaller than cities, with most Pennsylvania cities having been incorporated as a borough before being incorporated as a city.[83] There are 958 boroughs in Pennsylvania, all of which governed by the "weak mayor" form of mayor–council government.[82] [83]

The skyline of Allentown, the third largest city in Pennsylvania.

Townships are the third type of municipality in Pennsylvania and are classified as either first class or second class townships. There are 1,454 second class townships and 93 first class townships.[86] Second class township can become first class townships if it has a population density greater than 300 inhabitants per square mile ( /km$^2$) and a referendum is passed supporting the change.[86]

There is one exception to the types of municipalities in Pennsylvania: Bloomsburg was incorporated as a town in 1870 and is, officially, the only town in the state.[87] In 1975, McCandless Township adopted a home-rule charter under the name of "Town of McCandless", but is, legally, still a first class township.[88]

The total of 56 cities, 958 boroughs, 93 first class townships, 1454 second class townships, and 1 town (Bloomsburg) is 2562 municipalities.

# Education

Pennsylvania has 500 public school districts, thousands of private schools, publicly funded colleges and universities, and over 100 private institutions of higher education.

The University of Pennsylvania quadrangle in Philadelphia

## Primary and secondary education

In general, under state law, school attendance in Pennsylvania is mandatory for a child from the age of 8 until the age of 17, or until graduation from an accredited high school, whichever is earlier.[89] As of 2005, 83.8% of Pennsylvania residents age 18 to 24 have completed high school. Among residents age 25 and over, 86.7% have graduated from high school. Additionally, 25.7% have gone on to obtain a bachelor's degree or higher.[90] State students consistently do well in standardized testing. In 2007, Pennsylvania ranked 14th in mathematics, 12th in reading, and 10th in writing for 8th grade students.[91]

In 1988, the Pennsylvania General Assembly passed Act 169, which allows parents or guardians to homeschool their children as an option for compulsory school attendance. This law specifies the requirements and responsibilities of the parents and the school district where the family lives.[92]

## Higher education

The Pennsylvania State System of Higher Education (PASSHE) is the public university system of the Commonwealth, with 14 state-owned schools. The Commonwealth System of Higher Education is organizing body of the 4 state-related schools in Pennsylvania, these schools are independent institutions that receive some state funding. There are also 15 publicly funded two-year community colleges and technical schools that are separate from the PASSHE system. Additionally there are many private two- and four-year technical schools, colleges and universities.

Carnegie Mellon University, The Pennsylvania State University, the University of Pennsylvania, and the University of Pittsburgh, are members of the Association of American Universities, an invitation only organization of leading research universities. The Pennsylvania State University is the Commonwealth's Land-grant university, Sea Grant College and, Space Grant College. The University of Pennsylvania, located in Philadelphia, is considered the first university in the United States and established the country's first medical school. The University of Pennsylvania is also the Commonwealth's only, and geographically the most southern, Ivy League school. The Pennsylvania Academy of the Fine Arts is the first and oldest art school in the United States.[] Philadelphia College of Pharmacy, now a part of University of the Sciences in Philadelphia, was the first pharmacy school in the United States.[]

# Recreation

Pennsylvania is home to the nation's first zoo, the Philadelphia Zoo.[93] Other long-accredited AZA zoos include the Erie Zoo and the Pittsburgh Zoo & PPG Aquarium. The Lehigh Valley Zoo and ZOOAMERICA are other notable zoos. The Commonwealth boasts some of the finest museums in the country, including the Carnegie Museums in Pittsburgh, the Philadelphia Museum of Art, and several others. One unique museum is the Houdini Museum in Scranton, the only building in the world devoted to the legendary magician.[94] Pennsylvania is also home to the National Aviary, located in Pittsburgh.

Dorney Park and Wildwater Kingdom's Steel Force and Thunderhawk roller coasters in Allentown

All 121 state parks in Pennsylvania feature free admission.

Pennsylvania offers a number of notable amusement parks, including Camel Beach, Conneaut Lake Park, Dorney Park & Wildwater Kingdom, Dutch Wonderland, DelGrosso Amusement Park, Hersheypark, Idlewild Park, Kennywood, Knoebels, Lakemont Park, Sandcastle Waterpark, Sesame Place, Great Wolf Lodge and Waldameer Park. Pennsylvania also is home to the largest indoor waterpark resort on the East Coast, Splash Lagoon in Erie.

Ohiopyle State Park timelapse

There are also notable music festivals that take place in Pennsylvania. These include Musikfest and NEARfest in Bethlehem, the Philadelphia Folk Festival, Creation Festival, the Great Allentown Fair, and Purple Door.

There are nearly one million licensed hunters in Pennsylvania. Whitetail deer, cottontail rabbits, squirrel, turkey, and grouse are common game species. Pennsylvania is considered one of the finest wild turkey hunting states in the Union, alongside Texas and Alabama. Sport hunting in Pennsylvania provides a massive boost for the Commonwealth's economy. A report from The Center for Rural Pennsylvania (a Legislative Agency of the Pennsylvania General Assembly) reported that hunting, fishing, and furtaking generated a total of $9.6 billion statewide.

The Boone and Crockett Club shows that five of the ten largest (skull size) black bear entries came from the state.[] The state also has a tied record for the largest hunter shot black bear in the Boone & Crockett books at 733 lb ( kg) and a skull of 23 3/16 tied with a bear shot in California in 1993.[] The largest bear ever found dead was in Utah in 1975, and the second largest was shot by a poacher in the state in 1987.[] Pennsylvania holds the second highest number of Boone & Crockett-recorded record black bears at 183, second only to Wisconsin's 299.[]

## Transportation

Exterior

Interior

The Pennsylvania Department of Transportation, abbreviated as PennDOT, owns 39861 miles ( km) of the 121770 miles ( km) of roadway in the state, making it the fifth largest state highway system in the United States.[95] The Pennsylvania Turnpike system is 535 miles ( km) long, with the mainline portion stretching from Ohio to Philadelphia and New Jersey.[95] It is overseen by the Pennsylvania Turnpike Commission. Another major east–west route is Interstate 80, whichs runs primarily in the northern tier of the state from Ohio to New Jersey at the Delaware Water Gap. Interstate 90 travels the relatively short distance between Ohio and New York through Erie County, in the extreme northwestern part of the state.

U.S. Route 220 as it passes through Lamar Township.

Primary north–south highways are Interstate 79 from its terminus in Erie through Pittsburgh to West Virginia, Interstate 81 from New York through Scranton, Lackawanna County and Harrisburg to Maryland and Interstate 476, which begins 7 miles ( km) north of the Delaware border, in Chester, Delaware County and travels 132 miles ( km) to Clarks Summit, Lackawanna County, where it joins I-81. All but 20 miles ( km) of I-476 is the Northeast Extension of the Pennsylvania Turnpike, while the highway south of the main line of the Pennsylvania Turnpike is officially called the "Veterans Memorial Highway", but is commonly referred to by locals as the "Blue Route".

The Southeastern Pennsylvania Transportation Authority (SEPTA) is the sixth largest transit agency in the United States and operates the commuter, heavy and light rail transit, and transit bus service in the Philadelphia metropolitan area. The Port Authority of Allegheny County is the 25th largest transit agency and provides transit bus and light rail service in and around Pittsburgh.[96]

Intercity passenger rail transit is provided by Amtrak, with the majority of traffic occurring on the *Keystone Service* in the high-speed Keystone Corridor between Harrisburg and Philadelphia's 30th Street Station before heading north to New York City; the *Pennsylvanian* follows the same route from New York City to Harrisburg, but extends out to Pittsburgh. The *Capitol Limited* also passes through Pittsburgh, as well as Connellsville, on its way from Chicago to Washington, D.C.[97] Traveling between Chicago and New York City, the *Lake Shore Limited* passes through Erie once in each direction.[97] There are 67 short-line, freight railroads operating in Pennsylvania, the highest number in any U.S. state.[97]

Pennsylvania has seven major airports: Philadelphia International, Pittsburgh International, Lehigh Valley International, Harrisburg International, Erie International, University Park Airport and Wilkes-Barre/Scranton International. A total of 134 public-use airport are located in the state.[97] The port of Pittsburgh is the second largest inland port in the United States and the 18th largest port overall; the Port of Philadelphia is the 24th largest port in the United States.[98] Pennsylvania's only port on the Great Lakes is located in Erie.

The Allegheny River Lock and Dam Two is the most-used lock operated by the United States Army Corps of Engineers of its 255 nationwide.[99] The dam impounds the Allegheny River near Downtown Pittsburgh.

## Sports

Further information: List of people from Pennsylvania

Pennsylvania is home to many professional sports teams; the Pittsburgh Steelers and Philadelphia Eagles of the National Football League, the Philadelphia Phillies and Pittsburgh Pirates of Major League Baseball, the Philadelphia 76ers of the National Basketball Association, the Philadelphia Flyers and Pittsburgh Penguins of the National Hockey League, the Philadelphia Union of Major League Soccer, the Erie Bayhawks of the National Basketball Association Development League, the Wilkes-Barre/Scranton Penguins and Hershey Bears of the American Hockey League, the Philadelphia Soul and the Pittsburgh Power of the Arena Football League, and the Philadelphia Wings of the National Lacrosse League. Among them, these teams have accumulated 7 World Series Championships (Pirates 5, Phillies 2), 16 National League Pennants (Pirates 9, Phillies 7), 3 pre-Super Bowl era NFL Championships (Eagles), 6 Super Bowl Championships (Steelers), 1 Arena Bowl Championship (Soul), 2 NBA Championships (76ers), 5 Stanley Cups (Penguins 3, Flyers 2), 11 Calder Cups (Bears), and 6 Champion's Cups (Wings).

There are many minor league baseball teams located throughout the state; several of these teams are associated with either the Phillies or the Pirates. In 2008, the Phillies moved their AAA-level team, the Lehigh Valley IronPigs, from Ottawa, Ontario, in Canada, to a newly constructed stadium, Coca-Cola Park in Allentown. The Lehigh Valley is a core fan base for both the Phillies and the Philadelphia Eagles, who conduct their pre-season training camp on the practice fields of Lehigh University. Therefore, expectations are that the Lehigh Valley IronPigs (named after pig iron, an instrumental part in the construction of steel which used to be a large part of the local economy for decades), is likely to prove popular among Allentown and Lehigh Valley Phillies fans.

The Phillies' AA team, also called the Phillies, is located in Reading, while the short-season A-level affiliate, called the Crosscutters, is located in Williamsport. The Pirates' AA team, the Curve, is located in Altoona. The short-season A-level affiliate, the State College Spikes, is located in State College. The Spikes share a stadium with the Penn State University baseball team. Other Major League Baseball teams have a presence in the state as well. The New York Yankees' AAA team, also called the Yankees, is located in Moosic, between Scranton and Wilkes-Barre in the northeastern part of the state. The Detroit Tigers' AA team, the SeaWolves, is located in Erie, and the Washington Nationals' AA affiliate, the Senators, plays in the capital of Harrisburg. Two independent-league teams, the Lancaster Barnstormers and York Revolution of the Atlantic League of Professional Baseball, are located in south-central Pennsylvania, while the Washington Wild Things of the Frontier League are located in the south-western corner of the state.

Each summer, the Little League World Series is held in South Williamsport, near where Little League Baseball was founded in Williamsport. Also, the first World Series between the Boston Pilgrims (which became the Boston Red Sox) and Pittsburgh Pirates was played in Pittsburgh in 1903.

Coca-Cola Park in Allentown, home of the Lehigh Valley IronPigs

College football is very popular in Pennsylvania. The Penn State University Nittany Lions were coached by Joe Paterno who has led Penn State to two national championships (1982 & 1986) as well as five undefeated seasons (1968, 1969, 1973, 1986 and 1994). On January 22, 2012 Joe Paterno passed away from lung cancer. Penn State now has a new coach, Bill O'Brien, who was serving as the offensive coordinator for the New England Patriots before becoming head coach of Penn State. Penn State plays its home games in the second largest stadium in the United States, Beaver Stadium, which seats 107,282. In addition, the University of Pittsburgh Panthers have won nine national championships (1915, 1916, 1918, 1929, 1931, 1934, 1936, 1937 and 1976) and have played eight undefeated seasons (1904, 1910, 1915, 1916, 1917, 1920, 1937 and 1976).[100] Pitt plays its home games at Heinz Field, a facility it shares with the Pittsburgh Steelers. Other Pennsylvania schools that have won national titles in football include Lafayette College (1896), Villanova University (FCS 2009), and the University of Pennsylvania (1895, 1897, 1904 and 1908).[101]

College basketball is also popular in the state, especially in the Philadelphia area where five universities, collectively termed the Big Five, have a rich tradition in NCAA Division I basketball. National titles in college basketball have been won by the following Pennsylvania universities: La Salle University (1954), Temple University (1938), University of Pennsylvania (1920 and 1921), University of Pittsburgh (1928 and 1930) and Villanova University (1985).[102] [103]

Soccer is gaining popularity within the state of Pennsylvania as well. With the addition of the Philadelphia Union in the MLS, the state now boasts three teams that are eligible to compete for the Lamar Hunt U.S. Open Cup annually. The other two teams are the Pittsburgh Riverhounds and the Harrisburg City Islanders, both of the United Soccer Leagues Second Division (USL-2). Within the American Soccer Pyramid, the MLS takes the first tier, while the USL-2 claims the third tier.

The Pocono 500 at Pocono Raceway in Long Pond

In motorsports, the Mario Andretti dynasty of race drivers hails from Nazareth in the Lehigh Valley. Notable Racetracks in Pennsylvania include the Jennerstown Speedway in Jennerstown, the Lake Erie Speedway in North East, the Mahoning Valley Speedway in Lehighton, the Motordome Speedway in Smithton, the Mountain Speedway in St. Johns, the Nazareth Speedway in Nazareth; and the Pocono Raceway in Long Pond, which is home to both the NASCAR sanctioned Pennsylvania 500 and Pocono 500 stock car races. The state is also home to Maple Grove Raceway, near Reading, which hosts major National Hot Rod Association sanctioned drag racing events each year.

There are also two motocross race tracks that host a round of the AMA Toyota Motocross Championships in Pennsylvania. High Point Raceway [104] in located in Mt. Morris, Pennsylvania, and Steel City is located in Delmont, Pennsylvania.

Horse racing courses in Pennsylvania consist of The Meadows near Pittsburgh, Pocono Downs in Wilkes-Barre, and Harrah's Philadelphia in Chester, which offer harness racing, and Penn National Race Course in Grantville, Parx Racing (formerly Philadelphia Park) in Bensalem, and Presque Isle Downs near Erie, which offer thoroughbred racing. Smarty Jones, the 2004 Kentucky Derby and Preakness Stakes winner, had Philadelphia Park as his home course.

Arnold Palmer, one of the 20th century's most notable pro golfers, comes from Latrobe, while Jim Furyk, a current PGA member, grew up near in Lancaster. PGA tournaments in Pennsylvania include the 84 Lumber Classic, played at Nemacolin Woodlands Resort, in Farmington and the Northeast Pennsylvania Classic, played at Glenmaura National Golf Club, in Moosic.

Philadelphia is home to LOVE Park, once a skateboarding mecca, and across from City Hall, host to ESPN's X Games in 2001 and 2002.[105]

## Food

In his book *Yo Mama Cooks Like a Yankee*, author Sharon Hernes Silverman calls Pennsylvania the snack food capital of the world.[106] It leads all other states in the manufacture of pretzels and potato chips. The Sturgis Pretzel House introduced the pretzel to America, and companies like Anderson Bakery Company, Intercourse Pretzel Factory, and Snyder's of Hanover are leading manufacturers in the Commonwealth. Two of the three companies that define the U.S. potato chip industry are based in Pennsylvania: Utz Quality Foods, Inc., which started making chips in Hanover, Pennsylvania in 1921, and Wise Snack Foods which started making chips in Berwick in 1921 (the third, Lay's Potato Chips, is a Texas company). Other companies such as Herr Foods, Martin's Potato Chips, Snyder's of Berlin (not associated with Snyder's of Hanover) and Troyer Farms Potato Products are popular chip manufacturers.

The U.S. chocolate industry is centered in Hershey, Pennsylvania, with Mars, Godiva, and Wilbur Chocolate Company nearby, and smaller manufacturers such as Asher's[107] in Souderton,[108] and Gertrude Hawk of Dunmore. Other notable companies include Just Born in Bethlehem, Pennsylvania, makers of Hot Tamales, Mike and Ikes, the Easter favorite marshmallow Peeps, and Boyer Brothers of Altoona, Pennsylvania, which is well known for its Mallo Cups. Auntie Anne's Pretzels began as a market-stand in Downingtown, Pennsylvania, and now has corporate headquarters in Lancaster City.[109] Traditional Pennsylvania Dutch foods include chicken potpie, ham potpie, schnitz un knepp (dried apples, ham, and dumplings), fasnachts (raised doughnuts), scrapple, pretzels, bologna, chow-chow, and Shoofly pie. Martin's Famous Pastry Shoppe, Inc., headquartered in Chambersburg, Pennsylvania, specializes in potato bread, another traditional Pennsylvania Dutch food. D.G. Yuengling & Son, America's oldest brewery, has been brewing beer in Pottsville since 1829.

Among the regional foods associated with Philadelphia are cheesesteaks, hoagie, soft pretzels, Italian water ice, scrapple, Tastykake, and strombolis. In Pittsburgh, tomato ketchup was improved by Henry John Heinz from 1876 to the early 20th century. Famous to a lesser extent than Heinz ketchup are the Pittsburgh's Primanti Brothers Restaurant sandwiches, pierogies, and city chicken. Outside of Scranton, in Old Forge there are dozens of Italian restaurants specializing in pizza made unique by thick, light crust and American cheese. Erie also has its share of unique foods, including Greek sauce and sponge candy. Sauerkraut along with pork and mashed potatoes is a common meal on New Year's Day in Pennsylvania.

## State symbols

- **Motto**: "Virtue, liberty, and independence"
- **Tree**: Eastern Hemlock[110]
- **Game bird**: Ruffed Grouse[111]
- **Flower**: Mountain Laurel[111]
- **Insect**: Pennsylvania Firefly[111]
- **Animal**: White-tailed deer[111]
- **Dog**: Great Dane[111]
- **Fish**: Brook Trout[111]
- **Fossil**: *Phacops rana*[110]
- **Beverage**: Milk[110]
- **Song**: "Pennsylvania"[112]
- **Ship**: US Brig *Niagara*[110]
- **Electric locomotive**: GG1 4859[110]
- **Steam locomotive**: K4s 1361 and K4s 3750[110]
- **Beautification plant**: Penngift Crownvetch[110]

The Ruffed Grouse

## Nicknames

US Brig *Niagara* in port

Pennsylvania has been known as the *Keystone State* since 1802,[113] based in part upon its central location among the original Thirteen Colonies forming the United States, and also in part because of the number of important American documents signed in the state (such as the Declaration of Independence). It was also a keystone state economically, having both the industry common to the North (making such wares as Conestoga wagons and rifles)[114] [115] and the agriculture common to the South (producing feed, fiber, food, and tobacco).[116]

Another one of Pennsylvania's nicknames is the *Quaker State*; in colonial times, it was known officially as the *Quaker Province*,[117] in recognition of Quaker[118] William Penn's *First Frame of Government*[119] constitution for Pennsylvania that guaranteed liberty of conscience. He knew of the hostility[120] Quakers faced when they opposed religious ritual, taking oaths, violence, war and military service, and what they viewed as ostentatious frippery.[121]

"**The Coal State**", "**The Oil State**", "**The Chocolate State**", and "**The Steel State**" were adopted when those were the state's greatest industries.[122]

"The State of Independence" currently appears on many road signs entering the state.

# Gallery

Aliquippa

Allentown

Altoona

Beaver Falls

Bedford

Bethlehem

Blairsville

Butler

Carlisle

Chambersburg

Clarion

Clearfield

Corry

Coudersport

Easton

Ebensburg

Edinboro

Erie

Franklin

Gettysburg

Greencastle

Greensburg

Hanover

Hazleton

Hershey

Hollidaysburg

Indiana

Jeannette

Johnstown

Kittanning

Lancaster

Latrobe

Lebanon

Lewisburg

Lewistown

Lock Haven

McKees Rocks

Meadville

Monessen

New Castle

New Hope

Palmyra

Philadelphia

Phoenixville

Pittsburgh

Pittston City

Pottstown

Pottsville

Punxsutawney

Reading

Renovo

Ridgway

Scranton

Selinsgrove

Shamokin

State College

Sunbury

Tyrone

Uniontown

Warren

Washington

Waynesboro

Wilkes-Barre

Williamsport

York

## See also

- Outline of Pennsylvania – organized list of topics about Pennsylvania
- Index of Pennsylvania-related articles

## Notes

[1] 2006 Statistical Abstract: Geography & Environment: Land and Land Use (http://www.census.gov/compendia/statab/tables/06s0347.xls)

[3] http://www.nws.noaa.gov/climate/xmacis.php?wfo=phi

[4] http://www.nws.noaa.gov/climate/xmacis.php?wfo=bgm

[5] http://www.nws.noaa.gov/climate/xmacis.php?wfo=ctp

[6] http://www.nws.noaa.gov/climate/xmacis.php?wfo=pbz

[7] http://www.nws.noaa.gov/climate/xmacis.php?wfo=cle

[25] Charter for the Province of Pennsylvania-1681 (http://avalon.law.yale.edu/17th_century/pa01.asp). This charter, granted by Charles II (England) to William Penn, constituted him and his heirs proprietors of the province, which, in honor of his father, Admiral William Penn, (whose cash advances and services were thus requited,) was called Pennsylvania. To perfect his title, William Penn purchased, on 1682-08-24, a quit-claim from the Duke of York to the lands west of the Delaware River embraced in his patent of 1664

[31] David Yount (2007). " *How the Quakers invented America* (http://books.google.com/books?id=pk7ycUq3cxsC&pg=PA82&dq&hl=en#v=onepage&q=&f=false)". Rowman & Littlefield. p.82. ISBN 0-7425-5833-9

[32] Sydney G. Fisher (2009). " *The Quaker Colonies* (http://books.google.com/books?id=zKzFgAlx1CkC&pg=PA13&dq&hl=en#v=onepage&q=&f=false)". Echo Library. p.13 ISBN 1-4068-5110-8

[33] Hamilton, Alexander and Syrett, Harold C. *The Papers of Alexander Hamilton*. 1963, page 240.

[41] Battle of Gettysburg (http://content.answers.com/main/content/wp/en/thumb/b/b3/200px-Dddr66.jpg)

[42] American FactFinder – Results (http://factfinder2.census.gov/bkmk/table/1.0/en/ACS/10_SF4/DP02/0400000US42)

[44] " Americans under age 1 now mostly minorities, but not in Ohio: Statistical Snapshot (http://www.cleveland.com/datacentral/index.ssf/2012/06/americas_under_age_1_populatio.html)". *The Plain Dealer*. June 3, 2012.

[48] http://www.census.gov/popest/states/asrh/tables/SC-EST2005-03-42.csv
[51] The Works of Voltaire, volume 19 (http://oll.libertyfund.org/Texts/Voltaire0265/Works/Vol19-HTMLs/0060-19_Pt02.html)
[52] Religious diversity in Pennsylvania (http://cip.cornell.edu/Dienst/UI/1.0/Summarize/psu.ph/1134140590)
[55] Appeals court races wrap up with focus on voter mobilization (http://www.phillyburbs.com/pb-dyn/news/103-11042003-190258.html)
[59] (http://www.deptofnumbers.com/unemployment/pennsylvania/)| title= Department of Numbers | accessdate=May 27, 2012
[60] Agricultural Census 2002 (http://www.nass.usda.gov/census/census02/profiles/pa/cp99042.PDF)
[61] (http://www.kyw1060.com/Pa--Lawmakers-Consider-Table-Games--More-Small-Cas/5763077)
[65] http://sites.state.pa.us/PA_Constitution.html
[66] Commonwealth of PA - Election Results 2012 (http://www.electionreturns.state.pa.us/ElectionsInformation.aspx?FunctionID=12&ElectionID=53)
[79] *Revenue Department Releases August Collections (09/01/2006)* (http://www.revenue.state.pa.us/revenue/CWP/view.asp?Q=261929&A=208). Retrieved September 25, 2006.
[82] *The Pennsylvania Manual*, p. 6-3.
[83] *Pennsylvania Manual*, p. 6-5.
[84] *The Pennsylvania Manual*, p. 6-46.
[86] *The Pennsylvania Manual*, p. 6-6.
[87] *The Pennsylvania Manual*, p. 6-22.
[88] Title 302, Pennsylvania Code, Section 23.1–101.
[89] (http://www.pde.state.pa.us/) 'Pennsylvania Department of Education (PDE). Retrieved on December 4, 2009.'
[90] (http://nces.ed.gov/) 'National Center for Education Statistics (NCES). Retrieved on December 4, 2009.'
[91] 'NCES.'
[92] (http://www.pde.state.pa.us/home_education/site/default.asp) 'Pennsylvania Department of Education: Home Education and Private Tutoring. Retrieved on December 4, 2009.'
[93] (http://www.philadelphiazoo.org/index.php?id=10_2_1)
[95] "Pennsylvania Department of Transportation Fact Book", p. 7.
[96] "2010 Public Transportation Fact Book", p. 8.
[97] "Pennsylvania Department of Transportation Fact Book", p. 10.
[98] Waterborne Commerce Statistics Center, p. 5-4.
[104] http://www.highpointmx.com/
[107] http://news.google.com/newspapers?id=hsQfAAAAIBAJ&sjid=i9gEAAAAIBAJ&pg=1978,4428241&dq=asher%27s-chocolates&hl=en
[110] *The Pennsylvania Manual*, p. 1-6
[111] *The Pennsylvania Manual*, p. 1-5
[112] *The Pennsylvania Manual*, p. 1-7
[113] PHMC: State Symbols (http://www.phmc.state.pa.us/bah/pahist/symbols.asp?secid=31)
[115] Lancaster rifle (http://www.2nj.org/library/weapons/flintlock_rifles.htm)
[116] PHMC: Agriculture in Pennsylvania (http://www.phmc.state.pa.us/ppet/agriculture/page1.asp?secid=31)
[117] The Quaker Province (http://www.dgs.state.pa.us/dgs/lib/dgs/pa_manual/section1/the_quaker_province_1681-1776.pdf#search="quaker province ")
[120] Pennsylvania translates to "Penn's Woods" and was named after the father of William Penn, the founder of the colony. Digital History: Persecution of the Quakers (http://www.digitalhistory.uh.edu/documents/documents_p2.cfm?doc=240)
[121] The Quaker Province 1681–1776 (http://www.phmc.state.pa.us/bah/pahist/quaker.asp?secid=31)

# External links

- Pennsylvania (http://www.dmoz.org/Regional/North_America/United_States/Pennsylvania/) at the Open Directory Project
- Gov. Andrew Curtin's Pennsylvania Reserve Volunteer Corps, Civil War 1861–1864 (http://www.pareserves.com/)
- Official state government site (http://www.pa.gov/)
- Pennsylvania Department of Transportation (http://www.dot.state.pa.us/)
- Allegheny National Forest (http://www.fs.fed.us/r9/forests/allegheny/)
- Pennsylvania Wilds (http://www.pawilds.com/)
- USGS real-time, geographic, and other scientific resources of Pennsylvania (http://www.usgs.gov/state/state.asp?State=PA)
- Energy Data & Statistics for Pennsylvania (http://tonto.eia.doe.gov/state/state_energy_profiles.cfm?sid=PA)

- Pennsylvania State Facts (http://www.ers.usda.gov/StateFacts/PA.htm)
- Official state tourism site (http://www.visitPA.com/)
- Biography of William Penn from 1829 (http://www.antiquebooks.net/readpage.html#penn)
- A History of Pennsylvania from 1905 (http://rack1.ul.cs.cmu.edu/is/pennhist/)
- Free Original Documents Online: Pennsylvania State Archives 1600s to 1800s (http://www.footnote.com/topicpage.php?tp=88)
- Pennsylvania Department of Community and Economic Development (http://newpa.com/)
- National Association of Counties (information on each Pennsylvania County)

# Lebanon,_Pennsylvania

**Lebanon**, formerly known as **Steitztown**,[1] is a city in and the county seat of Lebanon County, Pennsylvania, United States.[] The population was 25,477 at the 2010 census, a 4.2% increase from the 2000 count of 24,461. Lebanon is located in the central part of the Lebanon Valley, 26 miles ( km) east of Harrisburg and 29 miles ( km) west of Reading.

## History

Native tribes in the area of what is now Lebanon included the Shawnee, Susquehannock, Gawanese, Lenape (or Delaware), and Nanticoke peoples.[2] Lebanon was then settled in 1720 by European colonists, many with the family names of "Steitz" and "Light", along a creek that was then named "Steitz Creek". The Light patriarchs built a fort to protect against Indians and named it "Light's Fort" during this time. The town was laid out in 1753. The town was incorporated as a borough on February 20, 1821 and became a city on November 25, 1885. It adopted the commission form of government, consisting of four councilmen and a mayor. In 1900, 17,628 people lived in Lebanon; in 1910, 19,240 people lived there; in 1920, 24,643 people lived there; and in 1940, 27,206 people lived in Lebanon, Pennsylvania.

Lebanon bologna was first made here. Lebanon was formerly home to a major steel mill operated by Bethlehem Steel.

The Josiah Funck Mansion, Cornwall & Lebanon Railroad Station, Chestnut Street Log House, St. Lukes Episcopal Church, Salem Evangelical Lutheran Church, Reading Railroad Station and Tabor Reformed Church are listed on the National Register of Historic Places.[]

## Geography

Lebanon is located at (40.341529, -76.420853).[]

According to the United States Census Bureau, the city has a total area of 4.2 square miles ( $km^2$), all of it land.

Lebanon is bordered to the north and east by North Lebanon Township (4.5 mi), to the south and east by South Lebanon Township (3.22 mi), to the west by West Lebanon Township (1.07 mi), and to the south and west by North Cornwall Township (4.38 mi). The Quittapahilla Creek drains the city westward into the Susquehanna River via the Swatara Creek.

## Demographics

| Historical populations | | |
|---|---|---|
| Census | Pop. | %± |
| 1810 | 1434 | — |
| 1820 | 1437 | 0.2% |
| 1830 | 1826 | 27.1% |
| 1840 | 1860 | 1.9% |
| 1850 | 2184 | 17.4% |
| 1860 | 4449 | 103.7% |
| 1870 | 6727 | 51.2% |
| 1880 | 8778 | 30.5% |
| 1890 | 14664 | 67.1% |
| 1900 | 17628 | 20.2% |
| 1910 | 19240 | 9.1% |
| 1920 | 24643 | 28.1% |
| 1930 | 25561 | 3.7% |
| 1940 | 27206 | 6.4% |
| 1950 | 28156 | 3.5% |
| 1960 | 30045 | 6.7% |
| 1970 | 28572 | −4.9% |
| 1980 | 25711 | −10.0% |
| 1990 | 24800 | −3.5% |
| 2000 | 24461 | −1.4% |
| 2010 | 25477 | 4.2% |
| Sources:[3] [4] [5] | | |

As of the 2010 census, the city was 74.1% White, 5.9% Black or African American, 0.5% Native American, 1.1% Asian, and 3.2% were two or more races. 32.1% of the population were of Hispanic of Latino ancestry[6].

As of the census[] of 2000, there were 24,461 people, 10,266 households, and 6,056 families residing in the city. The population density was 5,844.8 people per square mile (2,254.0/km²). There were 11,220 housing units at an average density of 2,681.0 per square mile (1,033.9/km²). The racial makeup of the city was 85.50% White, 3.23% African American, 0.28% Native American, 1.02% Asian, 0.10% Pacific Islander, 8.11% from other races, and 1.76% from two or more races. Hispanic or Latino of any race were 16.43% of the population.

There were 10,266 households out of which 28.4% had children under the age of 18 living with them, 38.7% were married couples living together, 15.0% had a female householder with no husband present, and 41.0% were non-families. 35.4% of all households were made up of individuals and 15.3% had someone living alone who was 65 years of age or older. The average household size was 2.32 and the average family size was 3.00.

In the city the population was spread out with 25.0% under the age of 18, 8.4% from 18 to 24, 29.5% from 25 to 44, 20.5% from 45 to 64, and 16.6% who were 65 years of age or older. The median age was 36 years. For every 100 females there were 94.1 males. For every 100 females age 18 and over, there were 90.8 males.

The median income for a household in the city was $27,259, and the median income for a family was $34,045. Males had a median income of $26,957 versus $20,162 for females. The per capita income for the city was $15,584. About 12.8% of families and 16.2% of the population were below the poverty line, including 24.7% of those under age 18 and 10.5% of those age 65 or over.

# Education

Public education is provided by the Lebanon School District and Cornwall-Lebanon School District. Private institutions include Lebanon Catholic High School, New Covenant Christian School and Lebanon Christian Academy. All three private institutions have a varsity sports department and an elementary, junior high, and senior high. Students in Lebanon School District also may attend the Lebanon County Career and Technology Center (LCCTC). The city is home to Harrisburg Area Community College's Lebanon Campus.

# Culture

Lebanon is one of several Pennsylvania towns to drop or raise a unique item at midnight on New Year's Eve. Godshall's Quality Meats, owners of Weaver's Famous Lebanon Bologna, donates a huge (150-pound ( kg)) Lebanon bologna for the annual festivity. It is encased in a metal frame and suspended from a fire department ladder truck, and donated to a local rescue mission after the celebration.[7]

Lebanon's 12 ft, 150 pound New Year's Eve bologna

An infamous 1878 murder in Fort Indiantown Gap resulted in a trial of six defendants who all had blue eyes. They were given the moniker *the Blue-eyed Six* by a newspaper reporter who attended the trial, held in the Lebanon county courthouse. Five of the six defendants were hanged at the county jail. The trial received worldwide publicity and provided inspiration to Arthur Conan Doyle in writing the Sherlock Holmes short story *The Adventure of the Red-Headed League*.[8]

At one point in history the Lebanon County courthouse and jail became the home of the Lebanon Farmers Market. However, the popular farmers market returned to the original 30,000 square foot Market House location on South 8th street in 2003. This popular indoor market includes vendors such as Candy Rama, Keener Poultry, The Pretzel Twist and TTT Star Sushi. In addition to the vendors, customers can sit down and dine upstairs in Niko's Restaurant.

In December 2008, the TV show *Dirty Jobs*, hosted by Mike Rowe, visited the Seltzer's Smokehouse Meats to film production of Lebanon bologna. Later, the show featured the Wertz Candy Shop.[9]

In 2010, an independent film drama *Lebanon, PA* was made.[10] While the movie was set in Lebanon, all filming was done in other parts of Pennsylvania.

Lebanon, Pennsylvania, is named after the ancient Middle Eastern nation of Lebanon, which is commonly pronounced /ˈlɛbənɒn/, rhyming with the name "John." However, locals consistently pronounce the Pennsylvania city's name /ˈlɛbənən/ ("Leb-a-*nin*") and many shorten it to two syllables—"Leb-nin" or even "Lep-nin." The latter is particularly identified with Pennsylvania Dutch heritage. [11]

## Notable people

- Thomas Albert, composer
- Sam Bowie, NBA center
- Kerry Collins, NFL quarterback
- Bobby Gerhart, NASCAR/ARCA driver
- Jamie Lynn Gray, sport shooter and gold medalist at the 2012 Summer Olympics
- Jared Odrick, NFL defensive end for the Miami Dolphins
- Frank Reich, former NFL quarterback
- George S. Rentz, World War II Navy Cross recipient and United States Navy Chaplain
- Dick Shiner, former NFL quarterback

## References

[1] Lebanon: About Lebanon (http://www.lebanonpa.org/lebanoncity/cwp/view.asp?a=3&Q=504332&PM=1&lebanoncityNav=)
[2] A Brief History of Lancaster County (http://web.archive.org/web/19990203143240/http://www.willowvalley.com/wvlancasterinfo.htm). Web.archive.org (1999-02-03). Retrieved on 2013-07-27.
[3] http://www.census.gov/prod/www/abs/decennial/1940.html
[4] http://www.census.gov/prod/www/abs/decennial/1960cenpopv1.html
[5] http://www.census.gov/prod/cen1990/cph2/cph-2-1-1.pdf
[6] http://www.usatoday.com/news/nation/census/profile/PA
[7] WHT.com news story (http://www.whtm.com/news/stories/1207/484193.html)
[8] Ludwig, Gary (1979). The Blue Eyed Six, A Historical Narrative. Lebanon, PA: Hodge Podge USA.
[9] TV.com episode recap (http://www.tv.com/dirty-jobs/fuel-tank-cleaner/episode/955684/recap.html)
[10] (http://lebanonpamovie.com/)
[11] http://dailyapple.blogspot.com/2007/08/apple-260-wilkes-barre-pa.html

## External links

- Lebanon, PA (http://www.lebanoncountypages.com/lebanon-city-pa)
- 29917|The Community of Lebanon Association (http://www.lebanonpa.com/)
- Lebanon Valley Chamber of Commerce (http://www.lvchamber.org/)
- Lebanon Valley Council on the Arts (http://www.lebanonartscouncil.org/)
- Lebanon School District (http://www.lebanon.k12.pa.us/)
- Cedar Crest High School (http://www.clsd.k12.pa.us/hs/)
- Lebanon County Career and Technology Center (http://www.lcctc.k12.pa.us/)
- Lebanon Christian Academy (http://www.odbcministries.org/lcahome.html/)
- Godshalls Quality Meats (New Year's Eve Bologna Drop) (http://www.godshalls.com)

# United_States_Census_Bureau

| Bureau of the Census | |
|---|---|
| Seal | |
| Logo | |
| **Agency overview** | |
| **Formed** | July 1, 1903 |
| **Preceding Agency** | Temporary census offices |
| **Headquarters** | Suitland, Maryland |
| **Employees** | 5,593 (2006) |
| **Annual budget** | US$3.1 billion (2009)<br>US$7.2 billion (est. 2010)<br>US$1.3 billion (est. 2011) |
| **Agency executive** | Director, Thomas L. Mesenbourg (acting) |
| **Parent agency** | Economics and Statistics Administration |
| **Website** | |
| www.census.gov [1] | |

The **United States Census Bureau** (officially the **Bureau of the Census**, as defined in Title 13 U.S.C. § 11 [2]) is a principal agency of the U.S. Federal Statistical System responsible for producing data about the American people and economy.

The primary mission of the Census Bureau is conducting the U.S. Census every ten years, which allocates the seats of the U.S. House of Representatives to the states based on their population.[3] In addition to the decennial census, the Census Bureau continually conducts dozens of other censuses and surveys, including the American Community Survey, the U.S. Economic Census, and the Current Population Survey.[3] Furthermore, economic and foreign trade indicators released by the federal government typically contain data produced by the Census Bureau. The various censuses and surveys conducted by the Census Bureau help allocate over $400 billion in federal funds every year and help states, local communities, and businesses make informed decisions.[]

The Census Bureau is situated within the U.S. Department of Commerce and its director is appointed by the President of the United States.

## Legal mandate

The Constitution of the United States (Article I, section II) directs that the population be enumerated at least once every ten years and the resulting counts used to set the number of members from each state in the House of Representatives and, by extension, in the Electoral College. The Census Bureau now conducts a full population count every 10 years in years ending with a 0 (zero) and uses the term "decennial" to describe the operation. Between censuses, the Census Bureau makes population estimates and projections.[4]

In addition, Census data directly affects how more than $400 billion per year in federal and state funding is allocated to communities for neighborhood improvements, public health, education, transportation and much more. The Census Bureau is mandated with fulfilling these obligations: the collecting of statistics about the nation, its people, and economy. The Census Bureau's legal authority is codified in Title 13 of the United States Code.

The Census Bureau also conducts surveys on behalf of various federal government and local government agencies on topics such as employment, crime, health, consumer expenditures, and housing. Within the bureau, these are known as "demographic surveys" and are conducted perpetually between and during decennial (10-year) population counts. The Census Bureau also conducts economic surveys of manufacturing, retail, service, and other establishments and of domestic governments.

Between 1790 and 1840, the census was taken by marshals of the judicial districts.[5] The Census Act of 1840 established a central office[6] which became known as the Census Office. Several acts followed that revised and authorized new censuses, typically at the 10-year intervals. In 1902 the temporary Census Office was moved under the Department of Interior, and in 1903 it was renamed the Census Bureau under the new Department of Commerce and Labor. The department was intended to consolidate overlapping statistical agencies, but Census Bureau officials were hindered by their subordinate role in the department.[7]

An act in 1920 changed the date and authorized manufacturing censuses every 2 years and agriculture censuses every 10 years.[8] In 1929, a bill was passed mandating that the House of Representatives be reapportioned based on the results of the 1930 Census.[8] In 1954, various acts were codified into Title 13 of the US Code.[9]

By law the Census Bureau must count everyone and submit state population totals to the U.S. President by December 31 of any year ending in a zero. States within the Union receive the results in the spring of the following year.

## Uses of census data

Many federal, state, local and tribal governments use census data to:

- Decide the location of new housing and public facilities,
- Examine the demographic characteristics of communities, states, and the USA,
- Plan transportation systems and roadways,
- Determine quotas and creation of police and fire precincts, and
- Create localized areas for elections, schools, utilities, etc.

## Data stewardship

The United States Census Bureau is committed to confidentiality, and guarantees non-disclosure of any addresses or personal information related to individuals or establishments. Title 13 of the U.S. Code establishes penalties for the disclosure of this information. All Census employees must sign a sworn affidavit of non-disclosure prior to employment.

The Bureau cannot share responses, addresses or personal information with anyone including United States or foreign government and law enforcement agencies such as the IRS or the FBI or Interpol. "Providing quality data, for public good—while respecting individual privacy and, at the same time, protecting confidentiality—is the Census Bureau's core responsibility", "Keeping the public's trust is critical to the Census's ability to carry out the mission as the leading source of quality data about the Nation's people and economy."[10] Only after 72 years does the information collected become available to other agencies or the general public.

In 1918, the Census Bureau released individual information regarding several hundred young men to the Justice Department and Selective Service system for the purpose of prosecutions for draft evasion.[11] [12] During World War II, the United States Census Bureau assisted the government's Japanese American internment efforts by providing confidential neighborhood information on Japanese-Americans. The Bureau's role was denied for decades but was finally proven in 2007.[13] [14]

## Organizational structure

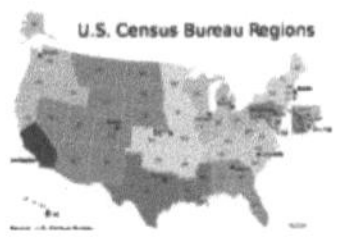

Pre-2013

Post-2013

Since 1903, the official census-taking agency of the United States government has been the Bureau of the Census. The Census Bureau is headed by a Director, assisted by a Deputy Director and an Executive Staff composed of the associate directors.

Census headquarters in Suitland, Maryland

The Census Bureau has had headquarters in Suitland, Maryland, since 1942. A new headquarters complex was completed in 2007 and supports over 4,000 employees.[15] The Bureau operates regional offices in 12 cities: Boston, New York City, Philadelphia, Detroit, Chicago, Kansas City, Seattle, Charlotte, Atlanta, Dallas, Denver, and Los Angeles. The National Processing Center is located in Jeffersonville, Indiana. Additional temporary processing facilities are used to facilitate the decennial census, which employs more than a million people. The cost of the 2000 Census was $4.5 billion. During the years just prior to the decennial census, parallel census offices, known as "Regional Census Centers" are opened in the field office cities. The decennial operations are carried out from these facilities. The Regional Census Centers oversee the openings and closings of smaller "Local Census Offices" within their collection jurisdictions. The estimated cost of the 2010 Census is $14.7 billion.

On January 1, 2013, the Census Bureau was to consolidate its 12 regional offices into 6. Increasing costs of data collection, changes in survey management tools such as laptops and the increasing use of multi-modal surveys (i.e. internet, telephone, and in-person) has led the Census Bureau to consolidate.[16] The remaining regional offices are to be located in: New York City, Philadelphia, Chicago, Atlanta, Denver, and Los Angeles.[17]

The Census Bureau also runs the Census Information Center cooperative program that involves 58 "national, regional, and local non-profit organizations." The CIC program aims to represent the interests of underserved communities.[18]

### Census regions and divisions

The United States Census Bureau has four official regions, with nine official divisions.[19]

- Northeast
  - New England
  - Middle Atlantic
- Midwest
  - East North Central States
  - West North Central States
- West
  - Pacific States
  - Mountain States
- South
  - West South Central States
  - East South Central States
  - South Atlantic States

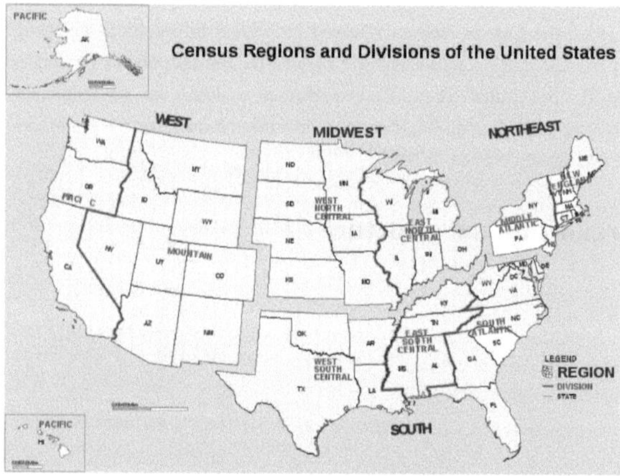


Map of official Census regions and divisions.

## Computer equipment

A card puncher, part of the tabulation system used to compile the thousands of facts gathered by the Bureau. Holes are punched in the card according to a prearranged code transferring the facts from the census questionnaire into statistics.

The 1890 census was the first to use the electric tabulating machines invented by Herman Hollerith.[20] [21] For 1890–1940 details, see In 1946, knowing of the Bureau's funding of Hollerith and, later, Powers, John Mauchly approached the Bureau about early funding for UNIVAC development.[22] A UNIVAC I computer was accepted by the Bureau in 1951.[23]

In order to reduce paper usage and reduce payroll expenses, 500,000 handheld computers (HHC's) were used for the first time in 2009 during the address canvassing portion of the 2010 Decennial Census Project. Projected savings were estimated to be over $1 billion.[24]

## Ongoing surveys

A survey is a method of collecting and analyzing social, economic, and geographic data. It provides information about the conditions of the United States, states, and counties. Throughout the decade between censuses, the bureau is conducts surveys to produce a general view and comprehensive study of the United States' social and economic conditions.

Staff from the Current Surveys Program conduct ongoing and special surveys about people and their characteristics. A network of professional field representatives gathers information from a sample of households, responding to questions about employment, consumer expenditures, health, housing, and other topics. Surveys conducted between decades:

- American Community Survey
- American Housing Survey
- Consumer Expenditure Survey
- Census of Governments
- Current Population Survey
- Economic Census
- National Hospital Discharge Survey
- National Health Interview Survey
- National Hospital Ambulatory Medical Care Survey
- National Crime Victimization Survey
- National Ambulatory Medical Care Survey (NAMCS)
- National Nursing Home Survey
- Survey of Income and Program Participation
- Survey of Construction
- Survey of Market Absorption
- Survey of Program Dynamics
- National Longitudinal Survey
- National Survey of Fishing, Hunting, & Wildlife-Associated Recreation 2001
- American Housing Survey
- Residential Finance Survey
- National Epidemiologic Survey of Alcohol Related Conditions
- Annual Retail Trade Survey
- Annual Wholesale Trade Survey
- Annual and Quarterly Services Surveys

## Other surveys conducted

The Census Bureau collects information in many other surveys and provides the data to the survey sponsor for release. These sponsors include:

- Bureau of Justice Statistics (BJS)
- Bureau of Labor Statistics (BLS)
- Bureau of Transportation Statistics (BTS)
- Department of Housing and Urban Development (HUD)
- National Center for Education Statistics (NCES)
- National Center for Health Statistics (NCHS)
- National Science Foundation (NSF)
- Social Security Administration (SSA)

## See also

- Outline of the United States
- Index of United States-related articles
- Book:United States
- Geography
  - Human geography
    - Demography
- United States
- Demographics of the United States
  - United States Census Bureau
    - List of U.S. states and territories by population
    - List of metropolitan areas of the United States
    - List of United States cities by population
    - List of United States counties and county-equivalents
  - United States Office of Management and Budget
    - The OMB has defined 1098 statistical areas comprising 388 MSAs, 541 μSAs, and 169 CSAs
      - Primary statistical area – List of the 574 PSAs
        - Combined Statistical Area – List of the 169 CSAs
        - Core Based Statistical Area – List of the 929 CBSAs
          - Metropolitan Statistical Area – List of the 388 MSAs
          - Micropolitan Statistical Area – List of the 541 μSAs

The 29 US MDs

- Micropolitan Statistical Area

The 576 US μSAs

- United States urban area

List of United States urban areas

## Notes

[1] http://www.census.gov
[2] http://www.law.cornell.edu/uscode/13/1.html
[3] USCB DOC-D1026 QVC Manual 01/03/09
[5] History 1790 (http://www.census.gov/history/www/overview/012270.html). US Census Bureau.
[6] History 1840 (http://www.census.gov/history/www/overview/012275.html). US Census Bureau.
[7] History: 1900 Overview (http://www.census.gov/history/www/through_the_decades/overview/1900.html). US Census Bureau.
[8] History 1920 (http://www.census.gov/history/www/overview/012283.html). US Census Bureau.
[9] History 1954 (http://www.census.gov/history/www/overview/012286.html). US Census Bureau.
[15] Census.gov (http://www.census.gov/aboutus/who.html)
[20] Herman Hollerith (http://www.census.gov/history/www/notable_alumni/010944.html)
[21] History 1890 (http://www.census.gov/history/www/overview/012280.html) US Census Bureau.
[24] Govcomm.harris.com (http://www.govcomm.harris.com/solutions/products/000138.asp)

## References

- The original version of, and subsequent additions to, this article were adapted from U.S. Census Bureau (http://www.census.gov/acsd/www/history.html) text.
- Geographic Areas Reference Manual (http://www.census.gov/geo/www/garm.html) from the U.S. Census Bureau contains detailed explanations of geographic terms used in the census.
- The mission and operation of the U.S. Census Bureau are laid out in Title 13 of the United States Code.

## External links

- United States Census Bureau (http://www.census.gov/)
  - USCB population estimates (http://www.census.gov/popest/data/index.html)
  - USCB History (http://www.census.gov/history)
  - U.S. and World Population Clocks – POPClocks (http://www.census.gov/main/www/popclock.html)
  - Works by Bureau of the Census (http://www.gutenberg.org/author/United_States._Bureau_of_the_Census) at Project Gutenberg
  - USCB Meeting Notices and Rule Changes (http://thefederalregister.com/b.p/agency/Census_Bureau/) from The Federal Register RSS Feed (http://thefederalregister.com/rss/agency/Census_Bureau/)
  - Proposed and finalized federal regulations from the USCB (http://openregs.com/agencies/view/173/bureau_of_the_census)

# Census

A **census** is the procedure of systematically acquiring and recording information about the members of a given population. It is a regularly occurring and official count of a particular population.[1] The term is used mostly in connection with national population and housing censuses; other common censuses include agriculture, business, and traffic censuses. The United Nations defines the essential features of population and housing censuses as "individual enumeration, universality within a defined territory, simultaneity and defined periodicity", and recommends that population censuses be taken at least every 10 years. United Nations recommendations also cover census topics to be collected, official definitions, classifications and other useful information to coordinate international practice.[2] [3]

Census taker visits a family living in a caravan, Netherlands 1925

The word is of Latin origin; during the Roman Republic, the census was a list that kept track of all adult males fit for military service. The modern census is essential to international comparisons of any kind of statistics and censuses collect data on many attributes of the population, not just how many people there are, although population estimates remain an important function of the census.

The census can be contrasted with sampling in which information is obtained only from a subset of a population, sometimes as an intercensal estimate. Census data are commonly used for research, business marketing, and planning, as well as a baseline for sampling surveys. Census counts are necessary to adjust samples to be representative of a population by weighting them as is common in opinion polling. Similarly, stratification requires knowledge of the relative sizes of different population strata which are derived from census enumerations. In some countries, census data are used to apportion electoral representation (sometimes controversially – e.g., *Utah v. Evans*).

## Sampling

A census is often construed as the opposite of a sample as everyone is surveyed rather than a fraction. However, population census relies on a sampling frame to count the population. This is the only way to be sure that everyone has been included as otherwise those not responding would not be followed up and individuals could be missed. The fundamental premise of a census is that the population is not known and a new estimate is to be made by the analysis of primary data.

The use of a sampling frame is counterintuitive as it suggests that the population size is already known. However, a census is also used to collect attribute data on the individuals in the nation. This process of sampling marks the difference between historical census, which was a house to house process or the product of an imperial decree, and the modern statistical project.

The sampling frame used by census is almost always an address register. Thus it is not known if there is anyone resident or how many people there are in each household. Depending on the mode of enumeration, a form is sent to the householder, an enumerator calls, or administrative records for the dwelling are accessed. As a preliminary to the dispatch of forms, census workers will check any address problems on the ground.

While it may seem straightforward to use the postal service file for this purpose, this can be out of date and some dwellings may contain a number of independent households. A particular problem is what are termed 'communal establishments' which may be student residences, religious orders, homes for the elderly etc. As these are not easily enumerated by a single householder, they are often treated differently and visited by special teams of census workers to make sure they are classified appropriately.

## Residence definitions

Individuals are normally counted within households and information is typically collected about the household structure and the housing. For this reason international documents refer to censuses of population and housing. Normally the census response is made by a household, indicating details of individuals resident there.

An important aspect of census enumerations is determining which individuals can be counted from which cannot be counted. Broadly, three definitions can be used: *de facto* residence; *de jure* residence; and, permanent residence. This is important to consider individuals who have multiple or temporary addresses. Every person should be identified uniquely as resident in one place but where they happen to be on Census Day, their *de facto* residence, may not be the best place to count them. Where an individual uses services may be more useful and this is at their usual, or *de jure*, residence. An individual may be represented at a permanent address, perhaps a family home for students or long term migrants.

It is necessary to have a precise definition of residence to decide whether visitors to a country should be included in the population count. This is becoming more important as students travel abroad for education for a period of several years. Other groups causing problems of enumeration are new born babies, refugees, people away on holiday, people moving home around census day, and people without a fixed address.

People having second homes because of working in another part of the country or retaining a holiday cottage are difficult to fix at a particular address sometimes causing double counting or houses being mistakenly identified as vacant. Another problem is where people use a different address at different times e.g. students living at their place of education in term time but returning to a family home during vacations or children whose parents have separated who effectively have two family homes. Census enumeration has always been based on finding people where they live as there is no systematic alternative - any list you could use to find people is derived from census activities in the first place. Recent UN guidelines provide recommendation on enumerating such complex households.[4]

## Enumeration strategies

Historical censuses used crude enumeration assuming absolute accuracy. Modern approaches take into account the problems of overcount and undercount, and the coherence of census enumerations with other official sources of data.[5] This reflects a realist approach to measurement, acknowledging that under any definition of residence there is a true value of the population but this can never be measured with complete accuracy. An important aspect of the census process is to evaluate the quality of the data.[6]

Many countries use a post-enumeration survey to adjust the raw census counts.[7] This works in a similar manner to capture-recapture estimation for animal populations. In census circles this method is called dual system enumeration (DSE). A sample of households are visited by interviewers who record the details of the household as at census day. These data are then matched to census records and the number of people missed can be estimated by considering the number missed in the census or survey but counted in the other. This way counts can be adjusted for non-response varying between different demographic groups. An explanation using a fishing analogy can be found in "Trout, Catfish and Roach..."[8] which won an award from the Royal Statistical Society for excellence in official statistics in 2011.

Triple system enumeration has been proposed as an improvement as it would allow evaluation of the statistical dependence of pairs of sources. However, as the matching process is the most difficult aspect of census estimation this has never been implemented for a national enumeration. It would also be difficult to identify three different sources that were sufficiently different to make the triple system effort worthwhile. The DSE approach has another weakness in that it assumes there is no person counted twice (over count). In *de facto* residence definitions this would not be a problem but in *de jure* definitions individuals risk being recorded on more than one form leading to double counting. A particular problem here are students who often have a term time and family address.

Several countries have used a system which is known as short form/long form.[9] This is a sampling strategy which randomly chooses a proportion of people to send a more detailed questionnaire to (the long form). Everyone receives the short form questions. Thereby more data are collected but not imposing a burden on the whole population. This also reduces the burden on the statistical office. Indeed in the UK all residents were required to fill in the whole form but only a 10% sample were coded and analysed in detail, until 2001.[10] New technology means that all data are now scanned and processed. Recently there has been controversy in Canada about the cessation of the long form with the head, Munir Sheikh resigning.[11]

The use of alternative enumeration strategies is increasing[12] but these are not so simple as many people assume. The Netherlands has been most advanced in adopting a census using administrative data. This allows a simulated census to be conducted by linking several different administrative databases at an agreed time. Data can be matched and an overall enumeration established accounting for where the different sources are discrepant. A validation survey is still conducted in a similar way to the post enumeration survey employed in a traditional census.

Other countries which have a population register use this as a basis for all the census statistics needed by users. This is most common amongst Nordic countries but requires a large number of different registers to be combined including population, housing, employment and education. These registers are then combined and brought up to the standard of a statistical register by comparing the data in different sources and ensuring the quality is sufficient for official statistics to be produced.[13]

A recent innovation is the French instigation of a rolling census programme with different regions enumerated each year such that the whole country is completely enumerated every 5 years.

In Europe, in connection with the 2010 census round, a large number of countries adopted alternative census methodologies, often based on the combination of data from registers, surveys and other sources.[14]

## Technology

Censuses have evolved in their use of technology with the latest censuses, the 2010 round, using many new types of computing. In Brazil, handheld devices were used by enumerators to locate residences on the ground. In many countries, census returns could be made via the Internet as well as in paper form. DSE is facilitated by computer matching techniques which can be automated, such as propensity score matching. In the UK, all census formats are scanned and stored electronically before being destroyed, replacing the need for physical archives. The record linking to perform an administrative census would not be possible without large databases being stored on computer systems.

New technology is not without problems in its introduction. The US census had intended to use the handheld computers but cost escalated and this was abandoned, with the contract being sold to Brazil. Online response is a good idea but one of the functions of census is to make sure everyone is counted accurately. A system which allowed people to enter their address without verification would be open to abuse. Therefore households have to be verified on the ground, typically by an enumerator visit or post out. Paper forms are still necessary for those without access to Internet connections. It is also plausible that the hidden nature of an administrative census means that users are not engaged with the importance of contributing their data to official statistics.

## Uses of Census data

In the nineteenth century, the first censuses collected paper enumerations that had to be collated by hand so the statistical uses were very basic. The government owned the data and were able to publish statistics themselves on the state of the nation. Uses were to measure changes in the population and apportion representation. Population estimates could be compared to those of other countries.

By the beginning of the twentieth century, censuses were recording households and some indications of their employment. In some countries, census archives are released for public examination after many decades, allowing genealogists to track the ancestry of interested people. Archives provide a substantial historical record which may challenge established notions of tradition. It is also possible to understand the societal history through job titles and arrangements for the destitute and sick.

As governments assumed responsibility for schooling and welfare, large government departments made extensive use of census data. Actuarial estimates could be made to project populations and plan for provision in local government and regions. It was also possible for central government to allocate funding on the basis of census data.

Even into the mid twentieth century, census data was only directly accessible to large government departments. However, computers meant that tabulations could be used directly by university researchers, large businesses and local government offices. They could use the detail of the data to answer new questions and add to local and specialist knowledge.

Now, census data are published in a wide variety of formats to be accessible to business, all levels of governance, media, students and teachers, charities and researchers, and any citizen who is interested. Data can be represented visually or analysed in complex statistical models, to show the difference between certain areas, or to understand the association between different personal characteristics. Census data offer a unique insight into small areas and small demographic groups which sample data would be unable to capture with precision.

## Privacy

Although the census provides a useful way of obtaining statistical information about a population, such information can sometimes lead to abuses, political or otherwise, made possible by the linking of individuals' identities to anonymous census data.[15] This consideration is particularly important when individuals' census responses are made available in microdata form, but even aggregate-level data can result in privacy breaches when dealing with small areas and/or rare subpopulations.

For instance, when reporting data from a large city, it might be appropriate to give the average income for black males aged between 50 and 60. However, doing this for a town that only has two black males in this age group would be a breach of privacy because either of those persons, knowing his own income and the reported average, could determine the other man's income.

Typically, census data are processed to obscure such individual information. Some agencies do this by intentionally introducing small statistical errors to prevent the identification of individuals in marginal populations;[16] others swap variables for similar respondents. Whatever measures have been taken to reduce the privacy risk in census data, new technology in the form of better electronic analysis of data poses increasing challenges to the protection of sensitive individual information. This known as statistical disclosure control.

Another possibility is to present survey results by means of statistical models in the form of a multivariate distribution mixture.[17] The statistical information in the form of conditional distributions (histograms) can be derived interactively from the estimated mixture model without any further access to the original database. As the final product does not contain any protected microdata, the model based interactive software [18] can be distributed without any confidentiality concerns.

Another method is simply to release no data at all, except very large scale data directly to the central government. Different release strategies between government have led to an international project (IPUMS) to co-ordinate access to microdata and corresponding metadata. Such projects also promote standardising metadata by projects such as SDMX so that best use can be made of the minimal data available.

## Historical examples

### Egypt

Censuses in Egypt are said to have been taken during the early Pharaonic period in 3340 BCE and in 3056 BCE.

### Ancient Greece

There are several accounts of ancient Greek and Mesopotamian city states carrying out censuses.[19] The question of which is first is clouded by very different approaches: counting only men, counting a pile of rocks etc. but such censuses took place 1600 BCE and earlier.

### Ancient Israel

Censuses are mentioned in the Bible. God commands a flat tax to be paid with the census in Exodus 30:11-16 for the upkeep of the Tabernacle. The Book of Numbers is named after the counting of the Israelite population (in Numbers 1-4) according to the house of the Fathers after the exodus from Egypt. A second census was taken while the Israelite were camped in the plains of Moab, in Numbers 26.

King David performed a census that produced disastrous results (in Samuel 2 24 and Chronicles 1 21). His son, King Solomon, had all of the foreigners in Israel counted in Chronicles 2 2:17.

The Gospel of Luke records Jesus being born during a census in Luke 2.

## China

In 2 CE during the Han Dynasty, China held a census still considered by scholars to be quite accurate.[20] [21] The census is considered one of the world's earliest preserved censuses,[] and found 57.67 million people registered in 12.36 million households.[22] [23] The areas with the highest population densities were the Yellow River and Huai River valleys. Chengdu was the largest city in China, reaching a population of 282,147 people.[] Another census dates to AD 144, when 49.73 million people were recorded living in 9.94 million households.

## India

The oldest recorded census in India is thought to have occurred around 300 BCE during the reign of the Emperor Chandragupta Maurya under the leadership of Kautilya or Chanakya.[24]

## Rome

The word "census" originated in ancient Rome from the Latin word *censere* ("to estimate"). The census played a crucial role in the administration of the Roman Empire, as it was used to determine taxes. With few interruptions, it was usually carried out every five years.[25] It provided a register of citizens and their property from which their duties and privileges could be listed. It is said to have been instituted by the Roman king Servius Tullius in the 6th century BCE,[26] at which time the number of arms-bearing citizens was counted at around 80,000.[27]

## Rashidun and Umayyad Caliphates

In the Middle Ages, the Caliphate began conducting regular censuses soon after its formation, beginning with the one ordered by the second Rashidun caliph, Umar.[28]

## Medieval Europe

The Domesday Book was undertaken in 1086 CE by William I of England so that he could properly tax the land he had recently conquered in medieval Europe. In 1183 CE, a census was taken of the crusader Kingdom of Jerusalem, to ascertain the number of men and amount of money that could possibly be raised against an invasion by Saladin, sultan of Egypt and Syria.

## Inca Empire

In the 15th century, the Inca Empire had a unique way to record census information. The Incas did not have any written language but recorded information collected during censuses and other numeric information as well as non-numeric data on quipus, strings from llama or alpaca hair or cotton cords with numeric and other values encoded by knots in a base-10 positional system.

## Spanish empire

On May 25, 1577, King Philip II of Spain ordered by royal cédula the preparation of a general description of Spain's holdings in the Indies. Instructions and a questionnaire, issued in 1577 by the Office of the Cronista Mayor, were distributed to local officials in the Viceroyalties of New Spain and Peru to direct the gathering of information. The questionnaire, composed of fifty items, was designed to elicit basic information about the nature of the land and the life of its peoples. The replies, known as "relaciones geográficas," were written between 1579 and 1585 and were returned to the Cronista Mayor in Spain by the Council of the Indies.

## See also

- IBM and the Holocaust
- *Liber Censuum*
- Race and ethnicity in censuses
- Social research

## Notes

[2] United Nations (2008). Principles and Recommendations for Population and Housing Censuses (http://unstats.un.org/unsd/demographic/sources/census/docs/P&R_Rev2.pdf). Statistical Papers: Series M No. 67/Rev.2. p8. ISBN 978-92-1-161505-0.

[3] http://www.unece.org/fileadmin/DAM/stats/publications/CES_2010_Census_Recommendations_English.pdf

[6] Leo Breiman (1994). "The 1991 Census Adjustment: Undercount or Bad Data?", *Statistical Science* **9** (4), pp.458-475.

[7] World Population and Housing Census Programme (2010) Post Enumeration Surveys: Operational guidelines (http://unstats.un.org/unsd/demographic/standmeth/handbooks/Manual_PESen.pdf), United Nations Secretariat, Dept of Economic and Social Affairs, Statistics Division, Tech Report

[8] Benton, P. Trout, Catfish and Roach: The beginner's guide to census population estimates (http://www.ons.gov.uk/ons/guide-method/census/2011/the-2011-census/census-coverage-survey/trout--catfish-and-roach.pdf), Office for National Statistics, UK

[9] Other methods of census taking (http://www.ons.gov.uk/ons/guide-method/census/2011/what-is-a-census/other-methods-of-census-taking/index.html), Office for National Statistics, UK

[15] The Census and Privacy (http://www.epic.org/privacy/census/).

[18] http://ro.utia.cas.cz/census

[21] Twitchett, D., Loewe, M., and Fairbank, J.K. *Cambridge History of China: The Ch'in and Han Empires 221 B.C.-A.D. 220*. Cambridge University Press (1986), p. 240.

[22] Nishijima (1986), 595–596.

[23] H. Yoon (1985). "An early Chinese idea of a dynamic environmental cycle", *GeoJournal* **10** (2), 211-212.

[25] Scheidel, Walter (2009) *Rome and China: comparative perspectives on ancient world empires*. Oxford University Press, p. 28.

[26] Livy *Ab urbe condita* 1.42

[27] Livy *Ab urbe condita* 1.42, citing Fabius Pictor

## References


- Alterman, Hyman, (1969). *Counting People: The Census in History*. Harcourt, Brace & Company.
- Bielenstein, Hans, (1978). "Wang Mang, the restoration of the Han dynasty, and Later Han." In *The Cambridge History of China*, vol. 1, eds. Denis Twitchett and John K. Fairbank, p. 223-90, Cambridge: Cambridge University Press.
- Krüger, Stephen, (Fall 1991). "The Decennial Census", 19 *Western State University Law Review* 1 (http://wsulawreview.org/Volume19Articles.pdf); available at HeinOnline (http://heinonline.org/HOL/LandingPage?collection=journals&handle=hein.journals/wsulr19&div=6&id=&page=) (subscription required).
- Effects of UK 'Jedi' hoax on 2001 UK census from ONS (http://www.statistics.gov.uk/cci/nugget.asp?id=297).
- U.S. Census Press Release on 1930 Census (http://www.census.gov/Press-Release/www/releases/archives/miscellaneous/000507.html).
- U.S. Census Press Release on Soundex and WPA (http://www.archives.gov/publications/prologue/spring_2002_soundex_projects.html).

## External links

- My Free Census (http://www.myfreecensus.com/) Find your ancestors in the U.S. census, and International census, with these free online census records, census images, census indexes and census databases in our easy to use census directory.
- Census of Ireland 1911 (http://www.census.nationalarchives.ie/).
- Online Historical Population Reports Project (OHPR) (http://www.histpop.org).
- PR as a function of census management: comparative analysis of fifteen census experiences (http://www.tesionline.com/intl/thesis.jsp?idt=8384)

# Population_density

**Population density** (in agriculture **standing stock** and standing crop) is a measurement of population per unit area or unit volume. It is frequently applied to living organisms, and particularly to humans. It is a key geographic term.[1]

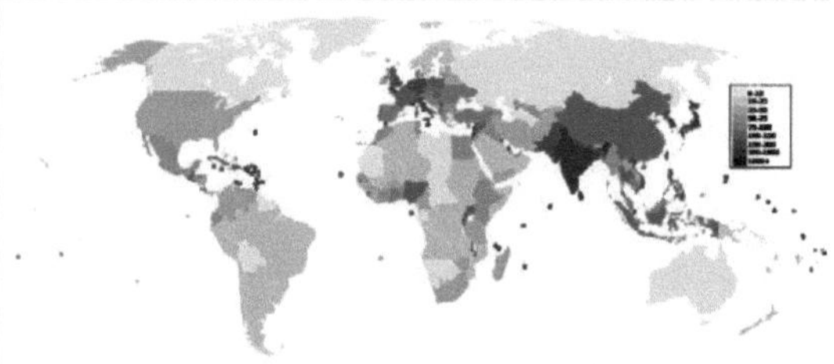

Population density (people per km$^2$) by country, 2006

## Biological population densities

Population density is population divided by total land area or water volume, as appropriate.[1]

Low densities may cause an extinction vortex and lead to further reduced fertility. This is called the Allee effect after the scientist who identified it. Examples of the causes in low population densities include:[2]

- Increased problems with locating sexual mates
- Increased inbreeding

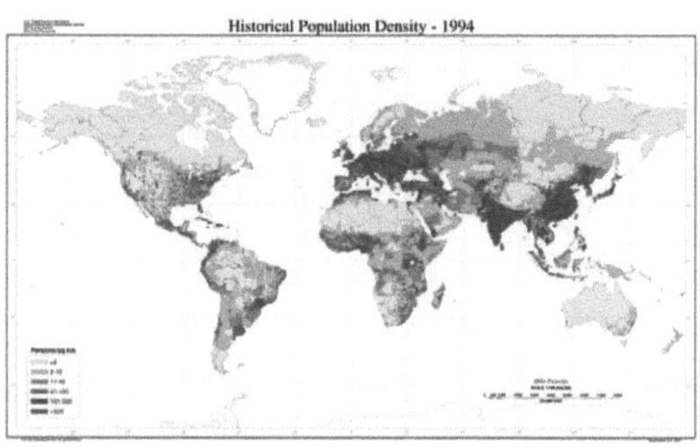


Population density (people per km$^2$) map of the world in 1994 (detailed).

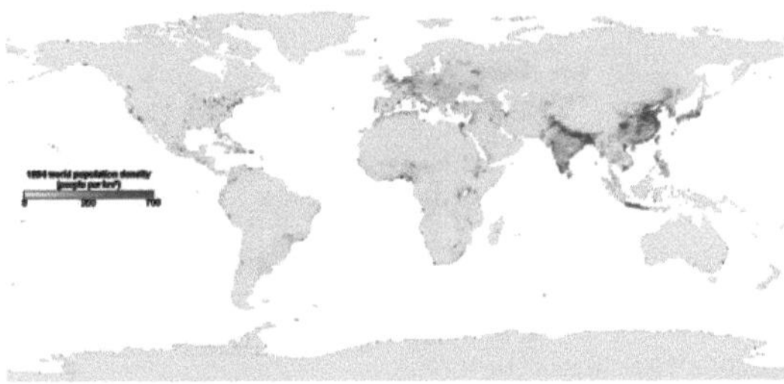

Population density (people per km$^2$) map of the world in 1994.

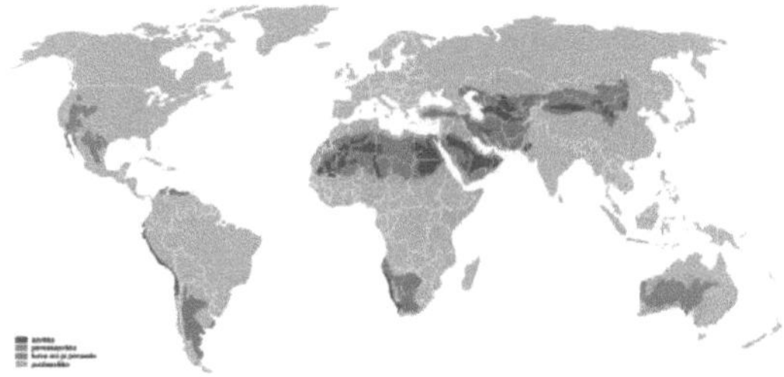

Deserts around the world. Compare with maps above.

## Human population density

Monaco in South Europe, currently holds the record for being the most densely populated nation in the world.

For humans, population density is the number of people per unit of area usually per square kilometer or mile (which may include or exclude cultivated or potentially productive area). Commonly this may be calculated for a county, city, country, another territory, or the entire world.

The world's population is 7 billion,[3] and Earth's total area (including land and water) is 510 million square kilometers (197 million square miles).[4] Therefore the worldwide human population density is 6.8 billion ÷ 510 million = 13.3 per km$^2$ (34.5 per sq. mile). If only the Earth's land area of 150 million km$^2$ (58 million sq. miles) is taken into account, then human population density increases to 45.3 per km$^2$ (117.2 per sq. mile). This calculation includes all continental and island land area, including Antarctica. If Antarctica is also excluded, then population density rises to 50 people per km$^2$ (129.28 per sq. mile).[1] Considering that over half of the Earth's land mass consists of areas inhospitable to human inhabitation, such as deserts and high mountains, and that population tends to cluster around seaports and fresh water sources, this number by itself does not give any meaningful measurement of human population density.

Several of the most densely populated territories in the world are city-states, microstates, or dependencies.[] [5] These territories share a relatively small area and a high urbanization level, with an economically specialized city population drawing also on rural resources outside the area, illustrating the difference between high population density and overpopulation.

Mongolia is the least densely populated country in the world.

Cities with high population densities are, by some, considered to be overpopulated, though the extent to which this is the case depends on factors like quality of housing and infrastructure and access to resources.[6] Most of the most densely populated cities are in southern and eastern Asia, though Cairo and Lagos in Africa also fall into this category.[7]

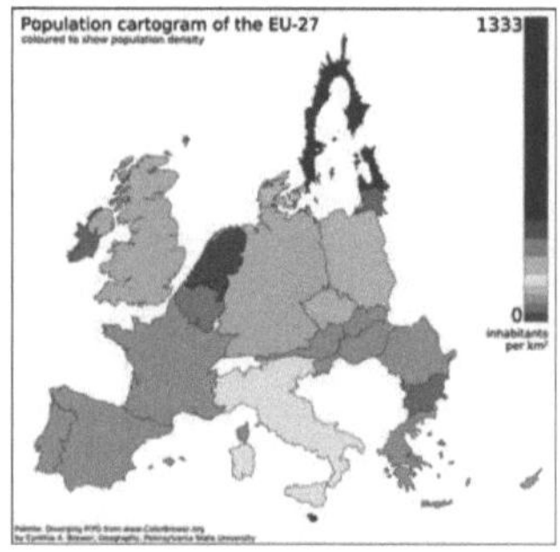


A population cartogram uses areas rather than colors to represent population.

City population is, however, heavily dependent on the definition of "urban area" used: densities are often higher for the central municipality itself, than when more recently developed and administratively separate suburban communities are included, as in the concepts of agglomeration or metropolitan area, the latter including sometimes neighboring cities. For instance, Milwaukee has a greater population density when just the inner city is measured, and not the surrounding suburbs as well.[8]

As a comparison, based on a world population of seven billion, the world's inhabitants would, as a loose crowd taking up ten square feet (one square metre) per person (Jacobs Method), would occupy a space a little larger than Delaware's land area.

## Most densely populated countries

### With population above 1 million

| Rank | Country/Region | Population | Area (km$^2$) | Density (Pop. per km$^2$) |
|---|---|---|---|---|
| 1 | Singapore | 5,183,700 | 710 | 7301 |
| 2 | Hong Kong | 7,061,200 | 1,104 | 6396 |
| 3 | Bahrain | 1,234,596 | 750 | 1646 |
| 4 | Bangladesh | 152,518,015 | 147,570 | 1034 |
| 5 | Taiwan (R.O.C) | 22,955,395 | 36,190 | 634 |
| 6 | Mauritius | 1,288,000 | 2,040 | 631 |
| 7 | South Korea | 48,456,369 | 99,538 | 487 |
| 8 | Rwanda | 10,718,379 | 26,338 | 407 |
| 9 | Netherlands | 16,760,000 | 41,526 | 404 |
| 10 | Lebanon | 4,224,000 | 10,452 | 404 |

### With population above 10 million

| Rank | Country/Region | Population | Area (km$^2$) | Density (Pop. per km$^2$) |
|---|---|---|---|---|
| 1 | Bangladesh | 152,518,015 | 147,570 | 1034 |
| 2 | Taiwan (R.O.C) | 22,955,395 | 36,190 | 634 |
| 3 | South Korea | 48,456,369 | 99,538 | 487 |
| 4 | Rwanda | 10,718,379 | 26,338 | 407 |
| 5 | Netherlands | 16,760,000 | 41,526 | 404 |
| 6 | India | 1,210,193,422 | 3,185,263 | 380 |
| 7 | Belgium | 11,007,020 | 30,528 | 361 |
| 8 | Japan | 127,960,000 | 377,944 | 339 |
| 9 | Sri Lanka | 20,653,000 | 65,610 | 345 |

## Other methods of measurement

While arithmetic density is the most common way of measuring population density, several other methods have been developed which aim to provide a more accurate measure of population density over a specific area.

- **Arithmetic density**: The total number of people / area of land (measured in km$^2$ or sq miles).
- **Physiological density**: The total population / area of arable land.
- **Agricultural density**: The total *rural* population / area of arable land.
- **Residential density** : The number of people living in an urban area / area of residential land.
- **Urban density** : The number of people inhabiting an urban area / total area of urban land.
- **Ecological optimum**: The density of population which can be supported by the natural resources.

## See also

- Demography
- Human geography
- Idealized population
- Optimum population
- Population bottleneck
- Population genetics
- Population health
- Population momentum
- Population pyramid
- Rural transport problem
- Small population size
- Distance sampling

**Lists:**

- List of cities by population
- List of cities by population density
- List of European cities proper by population density
- List of islands by population density
- List of countries by population density
- List of U.S. states by population density

## References

[1] Matt Rosenberg Population Density (http://geography.about.com/od/populationgeography/a/popdensity.htm). Geography.about.com. March 2 2011. Retrieved on 2011-12-10.
[2] Minimum viable population size (http://www.eoearth.org/article/Minimum_viable_population_size). Eoearth.org (2010-03-06). Retrieved on 2011-12-10.
[3] U.S. & World Population Clocks (http://www.census.gov/main/www/popclock.html). Census.gov. Retrieved on 2011-12-10.
[4] World (https://www.cia.gov/library/publications/the-world-factbook/geos/xx.html). CIA World Handbook
[5] The Monaco government uses a smaller surface area figure resulting in a population density of 18,078 per $km^2$
[6] Human Population (http://www.globalissues.org/issue/198/human-population). Global Issues. Retrieved on 2011-12-10.
[7] The largest cities in the world by land area, population and density (http://www.citymayors.com/statistics/largest-cities-density-125.html). Citymayors.com. Retrieved on 2011-12-10.
[8] The Population of Milwaukee County (http://www.wisconline.com/greenmap/milwaukee/population.html). Wisconline.com. Retrieved on 2011-12-10.

## External links

- Selected Current and Historic City, Ward & Neighborhood Densities (http://www.demographia.com/db-citydenshist.htm)

# Time_zone

A **time zone** is a region on Earth that has a uniform standard time for legal, commercial, and social purposes. It is convenient for areas in close commercial or other communication to keep the same time, so time zones tend to follow the boundaries of countries and their subdivisions.

Most of the time zones on land are offset from Coordinated Universal Time (UTC) by a whole number of hours (UTC−12 to UTC+14), but a few are offset by 30 or 45 minutes. Some higher latitude countries use daylight saving time for part of the year, typically by changing clocks by an hour. Many land time zones are skewed toward the west of the corresponding nautical time zones. This also creates a permanent daylight saving time effect.

## History

### Early timekeeping

Before the invention of clocks, people marked the time of day with apparent solar time (also called "true" solar time) – for example, the time on a sundial – which was typically different for every settlement.

When well-regulated mechanical clocks became widespread in the early 19th century, each city began to use some local mean solar time. Apparent and mean solar time can differ by up to around 15 minutes (as described by the equation of time) due to the non-circular shape of the Earth's orbit around the sun and the tilt of the Earth's axis. Mean solar time has days of equal length, and the difference between the two averages to zero after a year.

Greenwich Mean Time (GMT) was established in 1675 when the Royal Observatory was built as an aid to mariners to determine longitude at sea, providing a standard reference time when each city in England kept a different local time.

## Railroad time

Plaque commemorating the Railway General Time Convention of 1883 in North America

Local solar time became increasingly awkward as railways and telecommunications improved, because clocks differed between places by an amount corresponding to the difference in their geographical longitude, which varied by four minutes for every degree of longitude. The difference between New York and Boston is about two degrees or 8 minutes, the difference between Sydney and Melbourne, Australia, is about 7 degrees or 28 minutes. Bristol is 2°35′ W(est) of Greenwich (East London), so when it is noon in Bristol, it is about 10 minutes past noon in London.[1] The use of time zones smooths out these differences.

The first adaptation of a standard time in the world was established on December 1, 1847, in Great Britain by railway companies using GMT kept by portable chronometers. The first of these companies to adopt standard time was the Great Western Railway (GWR) in November 1840. This quickly became known as Railway Time. About August 23, 1852, time signals were first transmitted by telegraph from the Royal Observatory, Greenwich. Even though 98% of Great Britain's public clocks were using GMT by 1855, it was not made Britain's legal time until August 2, 1880. Some old British clocks from this period have two minute hands—one for the local time, one for GMT.[2]

The increase in worldwide communication had further increased the need for interacting parties to communicate mutually comprehensible time references to one another. The problem of differing local times could be solved across larger areas by synchronizing clocks worldwide, but in many places the local time would then differ markedly from the solar time to which people were accustomed. Time zones were a compromise, relaxing the complex geographic dependence while still allowing local time to approximate the mean solar time.

On November 2, 1868, the then-British colony of New Zealand officially adopted a standard time to be observed throughout the colony, and was perhaps the first country to do so. It was based on the longitude 172°30′ East of Greenwich, that is 11 hours 30 minutes ahead of GMT. This standard was known as New Zealand Mean Time.

Timekeeping on the American railroads in the mid-19th century was somewhat confused. Each railroad used its own standard time, usually based on the local time of its headquarters or most important terminus, and the railroad's train schedules were published using its own time. Some major railroad junctions served by several different railroads had a separate clock for each railroad, each showing a different time; the main station in Pittsburgh, Pennsylvania, for example, kept six different times.

Charles F. Dowd proposed a system of one-hour standard time zones for American railroads about 1863, although he published nothing on the matter at that time and did not consult railroad officials until 1869. In 1870, he proposed four ideal time zones (having north–south borders), the first centered on Washington, D.C., but by 1872 the first was centered 75°W of Greenwich, with geographic borders (for example, sections of the Appalachian Mountains). Dowd's system was never accepted by American railroads. Instead, U.S. and Canadian railroads implemented a version proposed by William F. Allen, the editor of the *Traveler's Official Railway Guide*.[3] The borders of its time zones ran through railroad stations, often in major cities. For example, the border between its Eastern and Central time zones ran through Detroit, Buffalo, Pittsburgh, Atlanta, and Charleston. It was inaugurated on Sunday, November 18, 1883, also called "The Day of Two Noons",[4] when each railroad station clock was reset as standard-time noon was reached within each time zone. The zones were named Intercolonial, Eastern, Central, Mountain, and Pacific. Within one year, 85% of all cities with populations over 10,000, about 200 cities, were using standard time.[5] A notable exception was Detroit (which is about half-way between the meridians of eastern time and central time), which kept local time until 1900, then tried Central Standard Time, local mean time, and Eastern Standard Time before a May 1915 ordinance settled on EST and was ratified by popular vote in August 1916. The confusion of times came to an end when Standard zone time was formally adopted by the U.S. Congress in the Standard Time Act of March 19, 1918.

### Worldwide time zones

Although the first person to propose a worldwide system of time zones was Italian mathematician Quirico Filopanti in his book *Miranda!* published in 1858, his idea was unknown outside the pages of his book until long after his death, so it did not influence the adoption of time zones during the 19th century. He proposed 24 hourly time zones, which he called "longitudinal days", the first centered on the meridian of Rome. He also proposed a universal time to be used in astronomy and telegraphy.[6] [7]

Scottish-born Canadian Sir Sandford Fleming proposed a worldwide system of time zones in 1879. He advocated his system at several international conferences, thus is widely credited with their invention. In 1876, his first proposal was for a global 24-hour clock, conceptually located at the center of the Earth and not linked to any surface meridian. In 1879 he specified that his universal day would begin at the anti-meridian of Greenwich (180th meridian), while conceding that hourly time zones might have some limited local use. He also proposed his system at the International Meridian Conference in October 1884, but it did not adopt his time zones because they were not within its purview. The conference did adopt a universal day of 24 hours beginning at Greenwich midnight, but specified that it "shall not interfere with the use of local or standard time where desirable".

By about 1900, almost all time on Earth was in the form of standard time zones, only some of which used an hourly offset from GMT. Many applied the time at a local astronomical observatory to an entire country, without any reference to GMT. It took many decades before all time on Earth was in the form of time zones referred to some "standard offset" from GMT/UTC. Most major countries had adopted hourly time zones by 1929. Nepal was the last country to adopt a standard offset, shifting slightly to UTC+5:45 in 1986.

Today, all nations use standard time zones for secular purposes, but they do not all apply the concept as originally conceived. Newfoundland, India, Iran, Afghanistan, Venezuela, Burma, the Marquesas, as well as parts of Australia use half-hour deviations from standard time, and some nations, such as Nepal, and some provinces, such as the Chatham Islands, use quarter-hour deviations. Some countries, most notably China and India, use a single time zone, even though the extent of their territory far exceeds 15° of longitude. Before 1949, China used five time zones (see Time in China).

## Definition

Before 1972, all time zones were specified as an offset from Greenwich Mean Time (GMT), which was the mean solar time at the meridian passing through the Royal Observatory in Greenwich, London. Since 1972 all official time services have broadcast radio time signals synchronized to UTC, a form of atomic time that includes leap seconds to keep it within 0.9 seconds of this former GMT, now called UT1. Many countries now legally define their standard time relative to UTC, although some still legally refer to GMT, including the United Kingdom itself. UTC, also called Zulu time, is used everywhere on Earth by astronomers and others who need to state the time of an event unambiguously.

Time zones are based on Greenwich Mean Time (GMT), the mean solar time at longitude 0° (the Prime Meridian). The definition of GMT was recently changed – it was previously the same as UT1, a mean solar time calculated directly from the rotation of the Earth. As the rate of rotation of the Earth is not constant, the time derived from atomic clocks was adjusted to closely match UT1. In January 1972, however, the length of the second in both Greenwich Mean Time and atomic time was equalized. The readings of participating atomic clocks are averaged out to give a uniform time scale.

Because the length of the average day is a small fraction of a second more than 24 hours (slightly more than 86400 seconds), leap seconds are periodically inserted into Greenwich Mean Time to make it approximate to UT1. This new time system is also called Coordinated Universal Time (UTC). Leap seconds are inserted to keep UTC within 0.9 seconds of UT1. Because the Earth's rotation is gradually slowing, leap seconds will need to be added more frequently in the future. However, from one year to the next the rotation rate is slightly irregular, so leap seconds are not added unless observations of Earth's rotation show that one is required. In this way, local times will continue to stay close to mean solar time and the effects of variations in Earth's rotation rate will be confined to simple step changes relative to the uniform time scale (International Atomic Time or TAI). All local times differ from TAI by an integral number of seconds. With the implementation of UTC, nations began to use it in the definition of their time zones. As of 2005, most nations had altered the definition of local time in this way.

In the UK, this involved redefining Greenwich Mean Time to make it the same as UTC.[8] British Summer Time (BST) is still one hour in advance of Greenwich Mean Time and is therefore also one hour in advance of Coordinated Universal Time. Thus Greenwich Mean Time is the local time at the Royal Observatory, Greenwich between 0100 hours GMT on the last Sunday in October and 0100 hours GMT on the last Sunday in March. Similar circumstances apply in many other places.

Leap seconds are considered by many to be a nuisance, and ways to abolish them are being considered. This means letting the time difference accumulate. One suggestion is to insert a "leap-hour" in about 5,000 years. For more on this discussion read Proposal to abolish leap seconds.

## Notation of time

### ISO 8601

#### UTC

If the time is in Coordinated Universal Time (UTC), add a "Z" directly after the time without a space. "Z" is the zone designator for the zero UTC offset. "09:30 UTC" is therefore represented as "09:30Z" or "0930Z". "14:45:15 UTC" would be "14:45:15Z" or "144515Z".

UTC time is also known as "Zulu" time, since "Zulu" is the ICAO spelling alphabet code word for "Z".

#### Offsets from UTC

Offsets from UTC are written in the format ±[hh]:[mm], ±[hh][mm], or ±[hh]. So if the time being described is one hour ahead of UTC (such as the time in Berlin during the winter), the zone designator would be "+01:00", "+0100", or simply "+01". This is appended to the time in the same way that 'Z' was above. The offset from UTC changes with daylight saving time, e.g. a time offset in Chicago, which is in the North American Central Time Zone, would be "−06:00" for the winter (Central Standard Time) and "−05:00" for the summer.

### Abbreviations

Time zones are often represented by abbreviations such as "EST, WST, CST" but these are not part of the international time and date standard ISO 8601 and their use as sole designator for a time zone is not recommended. Such designations can be ambiguous. For example, "BST", which is British Summer Time, was renamed "British Standard Time" between 1968 and 1971 when Central European Time was in force because legislators objected to calling it Central European Time. The same legislation affirmed that the Standard Time within the United Kingdom was, and would continue to be, Greenwich Mean Time.

## UTC offsets worldwide

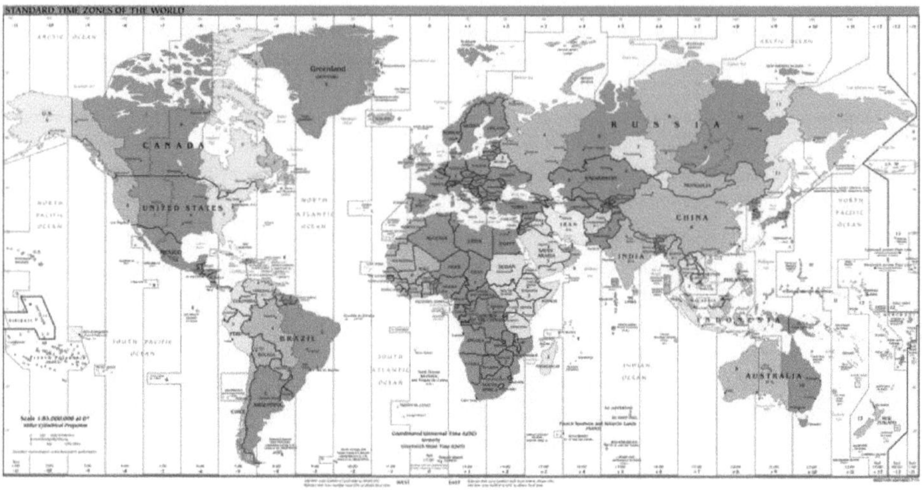


Time zones map (Last update: March 2013)

## List of UTC offsets

These examples give the local time at various locations around the world when daylight saving time is not in effect:

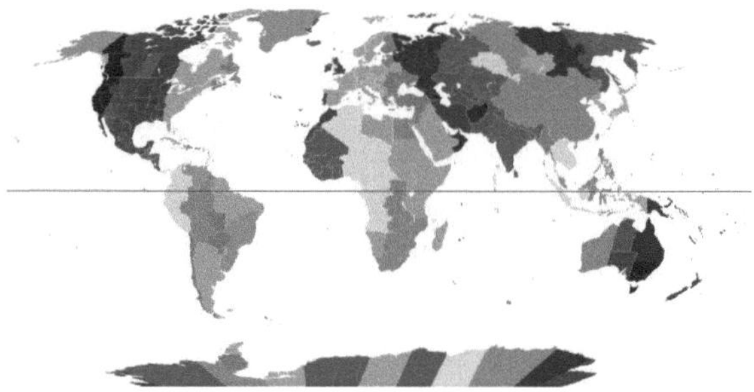

XX = ISO 3166-1 alpha-2 country code, XX- = parts of the country, N = North, S = South, UTC = Universal Coordinated Time, DST = Daylight Saving Time

| Time zone | Example time (ISO 8601 notation) | Example locations | Example locations that in summer use DST |
|---|---|---|---|
| UTC−12:00 | 2013-08-12T20:01:57-12:00 | Baker Island, Howland Island (both uninhabited) | |
| UTC−11:00 | 2013-08-12T21:01:57-11:00 | American Samoa, Niue | |
| UTC−10:00 | 2013-08-12T22:01:57-10:00 | United States (Hawaii) | Most of French Polynesia, United States (Aleutian Islands) |
| UTC−09:30 | 2013-08-12T22:31:57-09:30 | Marquesas Islands | |
| UTC−09:00 | 2013-08-12T23:01:57-09:00 | Gambier Islands | United States (most of Alaska) |
| UTC−08:00 | 2013-08-13T00:01:57-08:00 | | Canada (most of British Columbia), Mexico (Baja California), United States (California, most of Nevada, most of Oregon, Washington (state)) |
| UTC−07:00 | 2013-08-13T01:01:57-07:00 | Canada (northeastern British Columbia), Mexico (Sonora), United States (Arizona) | Canada (Alberta), Mexico (Chihuahua), United States (Colorado) |
| UTC−06:00 | 2013-08-13T02:01:57-06:00 | Canada (almost all of Saskatchewan), Costa Rica, El Salvador, Ecuador (Galápagos Islands), Guatemala, Honduras, Mexico (most), Nicaragua, | Canada (Manitoba), United States (Illinois, most of Texas) |
| UTC−05:00 | 2013-08-13T03:01:57-05:00 | Colombia, Cuba, Ecuador (continental), Jamaica, Panama, Peru | Canada (most of Ontario, most of Quebec), Haiti, United States (most of Florida, Georgia, Massachusetts, most of Michigan, most of Indiana, New York, North Carolina, Ohio, Washington, D.C.) |
| UTC−04:30 | 2013-08-13T03:31:57-04:30 | Venezuela | |
| UTC−04:00 | 2013-08-13T04:01:57-04:00 | Bolivia, Brazil (Amazonas), Chile (continental), Dominican Republic, Canada (Nova Scotia), Puerto Rico, Trinidad and Tobago | Falkland Islands |
| UTC−03:30 | 2013-08-13T04:31:57-03:30 | | Canada (island of Newfoundland and southern Labrador) |
| UTC−03:00 | 2013-08-13T05:01:57-03:00 | Argentina, Paraguay | Brazil (Federal District, Rio de Janeiro, São Paulo), most of Greenland, Uruguay |
| UTC−02:00 | 2013-08-13T06:01:57-02:00 | Brazil (Fernando de Noronha), South Georgia and the South Sandwich Islands | |
| UTC−01:00 | 2013-08-13T07:01:57-01:00 | Portugal (Azores), Cape Verde | |
| UTC±00:00 | 2013-08-13T08:01:57+00:00 | Côte d'Ivoire, Faroe Islands, Ghana, Iceland, Senegal | Ireland, Portugal (continental and Madeira), Spain (Canary Islands), Morocco, United Kingdom, Western Sahara |
| UTC+01:00 | 2013-08-13T09:01:57+01:00 | Algeria, Angola, Benin, Cameroon, Gabon, Niger, Nigeria, Democratic Republic of the Congo (west), Tunisia | Albania, Austria, Belgium, Bosnia and Herzegovina, Spain (continental), Croatia, Czech Republic, Denmark, Germany, Hungary, Italy, Libya, Malta, Kosovo, Macedonia, France (metropolitan), Namibia, the Netherlands, Norway, Poland, Serbia, Slovakia, Slovenia, Sweden, Switzerland |
| UTC+02:00 | 2013-08-13T10:01:57+02:00 | Egypt, Malawi, Mozambique, South Africa, Swaziland, Zambia, Zimbabwe | Bulgaria, Cyprus, Estonia, Finland, Greece, Israel, Latvia, Lebanon, Lithuania, Moldova, Northern Cyprus, Palestine, Romania, Syria, Transnistria, Turkey, Ukraine |

| | | | |
|---|---|---|---|
| UTC+03:00 | 2013-08-13T11:01:57+03:00 | Belarus, Djibouti, Eritrea, Ethiopia, Iraq, Jordan, Kenya, Madagascar, Russia (Kaliningrad Oblast), Saudi Arabia, South Sudan, Sudan, Somalia, Somaliland, Tanzania, Uganda, Yemen | |
| UTC+03:30 | 2013-08-13T11:31:57+03:30 | | Iran |
| UTC+04:00 | 2013-08-13T12:01:57+04:00 | Abkhazia, Armenia, Azerbaijan, Georgia, Mauritius, Oman, Russia (European), Seychelles, South Ossetia, United Arab Emirates | |
| UTC+04:30 | 2013-08-13T12:31:57+04:30 | Afghanistan | |
| UTC+05:00 | 2013-08-13T13:01:57+05:00 | Kazakhstan (west), Maldives, Pakistan, Uzbekistan | |
| UTC+05:30 | 2013-08-13T13:31:57+05:30 | India, Sri Lanka | |
| UTC+05:45 | 2013-08-13T13:46:57+05:45 | Nepal | |
| UTC+06:00 | 2013-08-13T14:01:57+06:00 | Kazakhstan (most), Bangladesh, Bhutan, Russia (Ural: Sverdlovsk Oblast, Chelyabinsk Oblast) | |
| UTC+06:30 | 2013-08-13T14:31:57+06:30 | Cocos Islands, Myanmar | |
| UTC+07:00 | 2013-08-13T15:01:57+07:00 | Jakarta, Russia (Novosibirsk Oblast), Thailand, Vietnam | |
| UTC+08:00 | 2013-08-13T16:01:57+08:00 | China, Hong Kong, Russia (Krasnoyarsk Krai), Malaysia, Philippines, Singapore, Taiwan, most of Mongolia, Western Australia | |
| UTC+08:45 | 2013-08-13T16:46:57+08:45 | Australia (Eucla) | |
| UTC+09:00 | 2013-08-13T17:01:57+09:00 | East Timor, Russia (Irkutsk Oblast), Japan, North Korea, South Korea | |
| UTC+09:30 | 2013-08-13T17:31:57+09:30 | Australia (Northern Territory) | Australia (South Australia) |
| UTC+10:00 | 2013-08-13T18:01:57+10:00 | Russia (Zabaykalsky Krai), Papua New Guinea, Australia (Queensland) | Australia (New South Wales, Tasmania, Victoria) |
| UTC+10:30 | 2013-08-13T18:31:57+10:30 | | Lord Howe Island |
| UTC+11:00 | 2013-08-13T19:01:57+11:00 | New Caledonia, Russia (Primorsky Krai), Solomon Islands | |
| UTC+11:30 | 2013-08-13T19:31:57+11:30 | Norfolk Island | |
| UTC+12:00 | 2013-08-13T20:01:57+12:00 | Kiribati (Gilbert Islands), Fiji, Russia (Kamchatka Krai) | New Zealand |
| UTC+12:45 | 2013-08-13T20:46:57+12:45 | | Chatham Islands |
| UTC+13:00 | 2013-08-13T21:01:57+13:00 | Kiribati (Phoenix Islands), Tonga, Tokelau | Samoa |
| UTC+14:00 | 2013-08-13T22:01:57+14:00 | Kiribati (Line Islands) | |

Where the adjustment for time zones results in a time at the other side of midnight from UTC, then the date at the location is one day later or earlier.

Some examples when UTC is 23:00 on Monday when or where daylight saving time is not in effect:

- Cairo, Egypt: UTC+02; 01:00 on Tuesday
- Wellington, New Zealand: UTC+12; 11:00 on Tuesday

Some examples when UTC is 02:00 on Tuesday when or where daylight saving time is not in effect:

- Honolulu, Hawaii, United States: UTC−10; 16:00 on Monday
- Toronto, Ontario, Canada: UTC−05; 21:00 on Monday

The time-zone adjustment for a specific location may vary because of daylight saving time. For example New Zealand, which is usually UTC+12, observes a one-hour daylight saving time adjustment during the Southern Hemisphere summer, resulting in a local time of UTC+13.

## Time zone conversions

Conversion between time zones obeys the relationship

"time in zone A" − "UTC offset for zone A" = "time in zone B" − "UTC offset for zone B",

in which each side of the equation is equivalent to UTC. (The more familiar term "UTC offset" is used here rather than the term "zone designator" used by the standard.)

The conversion equation can be rearranged to

"time in zone B" = "time in zone A" − "UTC offset for zone A" + "UTC offset for zone B".

For example, what time is it in Los Angeles (UTC offset= −08) when the New York Stock Exchange opens at 09:30 (−05)?

time in Los Angeles = 09:30 − (−05:00) + (−08:00) = 06:30.

In Delhi (UTC offset= +5:30), the New York Stock Exchange opens at

time in Delhi = 09:30 − (−05:00) + (+5:30) = 20:00.

These calculations become more complicated near a daylight saving boundary (because the UTC offset for zone X is a function of the UTC time).

## Nautical time zones

Since the 1920s a nautical standard time system has been in operation for ships on the high seas. Nautical time zones are an ideal form of the terrestrial time zone system. Under the system, a time change of one hour is required for each change of longitude by 15°. The 15° gore that is offset from GMT or UT1 (not UTC) by twelve hours is bisected by the nautical date line into two 7.5° gores that differ from GMT by ±12 hours. A nautical date line is implied but not explicitly drawn on time zone maps. It follows the 180th meridian except where it is interrupted by territorial waters adjacent to land, forming gaps: it is a pole-to-pole dashed line.[9] [10] [11]

A ship within the territorial waters of any nation would use that nation's standard time, but would revert to nautical standard time upon leaving its territorial waters. The captain is permitted to change the ship's clocks at a time of the captain's choice following the ship's entry into another time zone. The captain often chooses midnight. Ships going in shuttle traffic over a time zone border often keeps the same time zone all the time, to avoid confusion about work, meal and shop opening hours. Still the time table for port calls must follow the land time zone.

## Skewing of zones

Difference between sun time and clock time during daylight saving time:

Ideal time zones, such as nautical time zones, are based on the mean solar time of a particular meridian located in the middle of that zone with boundaries located 7.5 degrees east and west of the meridian. In practice, zone boundaries are often drawn much farther to the west with often irregular boundaries, and some locations base their time on meridians located far to the east.

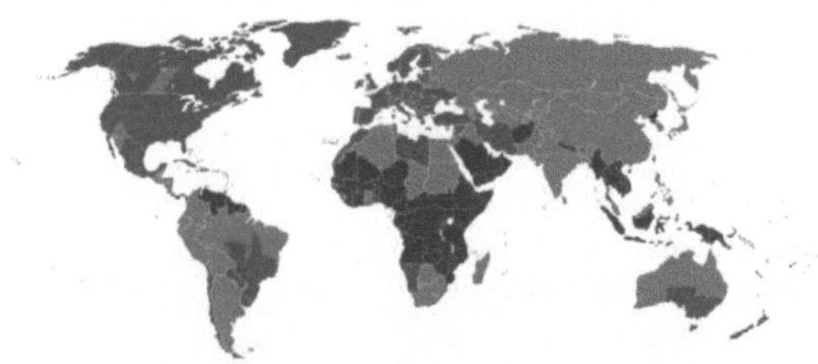

DST observed DST no longer observed DST never observed

For example, even though the Prime Meridian (0°) passes through Spain and France, they use the mean solar time of 15 degrees east (Central European Time) rather than 0 degrees (Greenwich Mean Time). France previously used GMT, but was switched to CET (Central European Time) during the German occupation of the country during World War II and did not switch back after the war.[12] Similarly, prior to World War II, the Netherlands observed "Amsterdam Time", which was twenty minutes ahead of Greenwich Mean Time. They were obliged to follow German time during the war, and kept it thereafter. In the mid 1970s the Netherlands, as with other European states, began observing daylight saving (summer) time.

There is a tendency to draw time zone boundaries far to the west of their meridians. Many of these locations also use daylight saving time. As a result, in the summer, solar noon in the Spanish town of Muxia occurs at 2:37 pm by the clock. This area of Spain never experiences sunset before 6:00 pm local time even in midwinter, despite its lying more than 40 degrees north of the equator. Near the summer solstice, Muxia has sunset times similar to those of Stockholm, which is in the same time zone and 16 degrees further north.

A more extreme example is Nome, Alaska, which is at 165°24′W longitude—just west of center of the idealized Samoa Time Zone (165°W). Nevertheless, Nome observes Alaska Time (135°W) with DST so it is slightly more than two hours ahead of the sun in winter and over three in summer.[13] Kotzebue, Alaska, also near the same meridian but north of the Arctic Circle, has an annual event on 9 August to celebrate *two* sunsets in the same 24-hour day, one shortly after midnight at the start of the day, and the other shortly before midnight at the end of the day.

Also, China extends as far west as 73°34′E, but all parts of it use UTC+08:00 (120°E), so solar "noon" can occur as late as 15:00 in western portions of China such as Xinjiang and Tibet.

## Daylight saving time

Many countries, and sometimes just certain regions of countries, adopt daylight saving time (also known as "Summer Time") during part of the year. This typically involves advancing clocks by an hour near the start of spring and adjusting back in autumn ("spring" forward, "fall" back). Modern DST was first proposed in 1907 and was in widespread use in 1916 as a wartime measure aimed at conserving coal. Despite controversy, many countries have used it off and on since then; details vary by location and change occasionally. Most countries around the equator do not observe daylight saving time, since the seasonal difference in sunlight is minimal.

## Additional information

- France has twelve time zones including those of Metropolitan France, French Guiana and numerous islands, inhabited and uninhabited. Russia has nine time zones (and used to have 11 time zones before March 2010), eight contiguous zones plus Kaliningrad exclave on the Baltic Sea. The United States has ten time zones (nine official plus that for Wake Island and its Antarctic stations; no official time zone is specified for uninhabited Howland Island and Baker Island). Australia has nine time zones (one unofficial and three official on the mainland plus four for its territories and one more for an Antarctic station not included in other time zones). The United Kingdom has eight time zones for itself and its overseas territories. Canada has six official time zones. The Danish Realm has five time zones.
- In terms of area, China is the largest country with only one time zone (UTC+08). China also has the widest spanning time zone. Before 1949, China was separated into five time zones.
- Stations in Antarctica generally keep the time of their supply bases, thus both the Amundsen-Scott South Pole Station (U.S.) and McMurdo Station (U.S.) use New Zealand time (UTC+12 southern winter, UTC+13 southern summer).
- The 27°N latitude passes back and forth across time zones in South Asia. Pakistan: +05:00, India +05:30, Nepal +05:45, +05:30, Myanmar +06:30. This switching was more odd in 2002, when Pakistan enabled daylight saving time. Thus from west to east, time zones were: +06:00, +05:30, +05:45, +05:30, +08:00, +06:00, +05:30 and +06:30.
- Because the earliest and latest time zones are 26 hours apart, any given calendar date exists at some point on the globe for 50 hours. For example, April 11 begins in time zone UTC+14 at 10:00 UTC April 10, and ends in time zone UTC−12 at 12:00 UTC April 12.
- There are 22 places where three or more time zones meet, for instance at the tri-country border of Finland, Norway and Russia. 28 countries present such triple points, with China being the most present (in 13 of the 22 triple points). Then come India (7), Russia, India and Afghanistan (4).
- There are 40 time zones. This is due to fractional hour offsets and zones with offsets larger than 12 hours near the International Date Line as well as one unofficial zone in Australia. See the list of time zones.
- The largest time gap along a political border is the 3.5 hour gap along the border of China (UTC+08) and Afghanistan (UTC+04:30).
- One of the most unusual time zones is the Australian Central Western Time zone (CWST), which is a small strip of Western Australia from the border of South Australia west to 125.5°E, just before Caiguna. It is 8¾ hours ahead of UTC (UTC+08:45) and covers an area of about 35,000 $km^2$, larger than Belgium, but has a population of about 200. Although unofficial, it is universally respected in the area—without it, the time gap in standard time at 129°E (the WA/SA border) would be 1.5 hours. See Time in Australia.

## Computer systems and the Internet

Computer operating systems include the necessary support for working with all (or almost all) possible local times based on the various time zones. Internally, operating systems typically use UTC as their basic time-keeping standard, while providing services for converting local times to and from UTC, and also the ability to automatically change local time conversions at the start and end of daylight saving time in the various time zones. (See the article on daylight saving time for more details on this aspect).

Web servers presenting web pages primarily for an audience in a single time zone or a limited range of time zones typically show times as a local time, perhaps with UTC time in brackets. More internationally-oriented websites may show times in UTC only or using an arbitrary time zone. For example the international English-language version of CNN includes GMT and Hong Kong Time,[14] whilst the US version shows Eastern Time.[15] US Eastern Time and Pacific Time are also used fairly commonly on many US-based English-language websites with global readership. The format is typically based in the W3C Note "datetime".

Email systems and other messaging systems (IRC chat, etc.)[16] time-stamp messages using UTC, or else include the sender's time zone as part of the message, allowing the receiving program to display the message's date and time of sending in the recipient's local time.

Database records that include a time stamp typically use UTC, especially when the database is part of a system that spans multiple time zones. The use of local time for time-stamping records is not recommended for time zones that implement daylight saving time due to the fact that once a year there is a one hour period when local times are ambiguous.

### Operating systems

#### Unix

Most Unix-like systems, including Linux and Mac OS X, keep system time as UTC (Coordinated Universal Time). Rather than having a single time zone set for the whole computer, timezone offsets can vary for different processes. Standard library routines are used to calculate the local time based on the current timezone, normally supplied to processes through the TZ environment variable. This allows users in multiple timezones to use the same computer, with their respective local times displayed correctly to each user. Time zone information most commonly comes from the IANA time zone database. In fact, many systems, including anything using the GNU C Library, can make use of this database.

#### Microsoft Windows

Windows-based computer systems prior to Windows 2000 used local time, but Windows 2000 and later can use UTC as the basic system time.[17] The system registry contains time zone information that includes the offset from UTC and rules that indicate the start and end dates for daylight saving in each zone. Interaction with the user normally uses local time, and application software is able to calculate the time in various zones. Terminal Servers allow remote computers to redirect their time zone settings to the Terminal Server so that users see the correct time for their time zone in their desktop/application sessions. Terminal Services uses the server base time on the Terminal Server and the client time zone information to calculate the time in the session.

## Programming languages

### Java

While most application software will use the underlying operating system for timezone information, the Java Platform, from version 1.3.1, has maintained its own timezone database. This database will need to be updated whenever timezone rules change. Sun provides an updater tool for this purpose.[18]

As an alternative to the timezone information bundled with the Java Platform, programmers may choose to use the Joda-Time library.[19] This library includes its own timezone data based on the IANA time zone database.[20]

### JavaScript

There is very little in the way of timezone support for JavaScript. Essentially the programmer has to extract the UTC offset by instantiating a time object, getting a GMT time from it, and differencing the two. This does not provide a solution for daylight saving variations.

### PHP

The DateTime objects and related functions have been compiled into the PHP core since 5.2. This includes the ability to get and set the default script timezone, and DateTime is aware of its own timezone internally. PHP.net provides extensive documentation on this.[21] As noted there, the most current timezone database can be implemented via the PECL timezonedb.

### Python

The standard module datetime stores and operates on the timezone information class tzinfo. The third party pytz module provides access to the full IANA time zone database.[22] Negated time zone offset in seconds is stored time.timezone and time.altzone attributes.

### Smalltalk

Each Smalltalk dialect comes with its own built-in classes for dates, times and timestamps, only a few of which implement the DateAndTime and Duration classes as specified by the ANSI Smalltalk Standard. VisualWorks provides a TimeZone class that supports up to two annually recurring offset transitions, which are assumed to apply to all years (same behavior as Windows time zones). Squeak provides a Timezone class that does not support any offset transitions. Dolphin Smalltalk does not support time zones at all.

For full support of the tz database (zoneinfo) in a Smalltalk application (including support for any number of annually recurring offset transitions, and support for different intra-year offset transition rules in different years) the third-party, open-source, ANSI-Smalltalk-compliant Chronos Date/Time Library is available for use with any of the following Smalltalk dialects: VisualWorks, Squeak, Gemstone, or Dolphin.[23]

## Databases

Some databases allow storage of a datetime type having time zone information. The SQL standard defines two standard time data types:

TIMESTAMP WITH TIME ZONE

TIMESTAMP WITHOUT TIME ZONE

However, the standard has a somewhat naive understanding of time zones. It generally assumes a time zone can be specified by a simple offset from GMT. This causes problems when trying to do arithmetic on dates which span daylight saving time transitions or which span political changes in time zone rules.

### Oracle

Oracle Database is configured with a database time zone, and connecting clients are configured with session time zones. Oracle Database uses two data types to store time zone information:

TIMESTAMP WITH TIME ZONE

Stores date and time information with the offset from UTC

TIMESTAMP WITH LOCAL TIME ZONE

Stores date and time information with respect to the dbtimezone (which cannot be changed so long as there is a column in the db of this type), automatically adjusting the date and time from the stored time zone to the client's session time zone.

### PostgreSQL

PostgreSQL uses the standard SQL data types but tries to impose an interpretation which avoids the problems described above.

TIMESTAMP WITH TIME ZONE

Stores date and time in UTC and converts to the client's local time zone (which could be different for each client) for display purposes and conversion to other types.

TIMESTAMP WITHOUT TIME ZONE

Stores date and time without any conversion on input or output. When converting to a TIMESTAMP WITH TIME ZONE, interprets it according to the client's local time zone.

TIME WITH TIME ZONE

Stores time of day together with a UTC offset in which it is to be interpreted.

TIME WITHOUT TIME ZONE

Stores time of day without any time zone specification.

## Time zones in outer space

Orbiting spacecraft typically experience many sunrises and sunsets in a 24-hour period, or in the case of Apollo program astronauts travelling to the moon, none. Thus it is not possible to calibrate time zones with respect to the sun, and still respect a 24-hour sleep/wake cycle. A common practice for space exploration is to use the Earth-based time zone of the launch site or mission control. This keeps the sleeping cycles of the crew and controllers in sync. The International Space Station normally uses Coordinated Universal Time (UTC).

Timekeeping on Mars can be more complex, since the planet has a solar day of approximately 24 hours and 39 minutes, known as a sol. Earth controllers for some Mars missions have synchronized their sleep/wake cycles with the Martian day, because solar-powered rover activity on the surface was tied to periods of light and dark. The difference in day length caused the sleep/wake cycles to slowly drift with respect to the day/night cycles on Earth, repeating approximately once every 36 days.

## See also

- Daylight saving time
- ISO 8601
- Jet lag
- List of time zone abbreviations
- List of time zones by country
- Lists of time zones
- Metric time
- Sunpendulum
- Time by country
- World clock

The control panel of the Time Zone Clock in front of Coventry Transport Museum.

## References

[1] Latitude and Longitude of World Cities http://www.infoplease.com/ipa/A0001769.html
[3] Economics of Time Zones (http://bpp.wharton.upenn.edu/mawhite/Papers/TimeZones.pdf) PDF (1.89 MB)
[6] Quirico Filopanti (http://www.scienzagiovane.unibo.it/english/scientists/filopanti-1.html) from scienzagiovane, Bologna University, Italy.
[7] Gianluigi Parmeggiani (Osservatorio Astronomico di Bologna), The origin of time zones (http://www.iav.it/planetario/didastro/didastro/english.htm)
[8] The Astronomical Almanac 1983, US Government Printing Office (Washington) and Her Majesty's Stationery Office (London), page B4.
[9] Bowditch, Nathaniel. *American Practical Navigator*. Washington: Government Printing Office, 1925, 1939, 1975.
[10] Hill, John C., Thomas F. Utegaard, Gerard Riordan. *Dutton's Navigation and Piloting*. Annapolis: United States Naval Institute, 1958.
[11] Howse, Derek. *Greenwich Time and the Discovery of the Longitude*. Oxford: Oxford University Press, 1980. ISBN 0-19-215948-8.
[17] GetSystemTime function (Windows) (http://msdn.microsoft.com/en-us/library/windows/desktop/ms724390(v=vs.85).aspx)
[23] Chronos Date/Time Library (http://www.chronos-st.org/)

## External links

- LEGAL TIME 2013 (http://www.itu.int/dms_pub/itu-t/opb/sp/T-SP-LT.1-2013-OAS-PDF-E.pdf)
- W3C Note Datetime (http://www.w3.org/TR/NOTE-datetime)
- US Official Time Clock (http://www.time.gov/timezone.cgi?UTC/s/0/java) Java-enabled clock to graphically display night and day around the globe.

# Per_capita_income

**Per capita income**, also known as income per person, is the mean income of the people in an economic unit such as a country or city. It is calculated by taking a measure of all sources of income in the aggregate (such as GDP or Gross national income) and dividing it by the total population.

## Per capita income as a measure of prosperity

Per capita income is often used as average income, a measure of the wealth of the population of a nation, particularly in comparison to other nations. Per capita income is often used to measure a country's standard of living. It is usually expressed in terms of a commonly used international currency such as the Euro or United States dollar, and is useful because it is widely known, easily calculated from readily-available GDP and population estimates, and produces a useful statistic for comparison of wealth between sovereign territories.this helps the country to know their development status.

Critics claim that per capita income has several weaknesses as an accurate measurement of prosperity:

- Comparisons of per capita income over time need to take into account changes in prices. Without using measures of income adjusted for inflation, they will tend to overstate the effects of economic growth.
- International comparisons can be distorted by differences in the costs of living between countries that aren't reflected in exchange rates. Where the objective of the comparison is to look at differences in living standards between countries, using a measure of per capita income adjusted for differences in purchasing power parity more accurately reflects the differences in what people are actually able to buy with their money.
- As it is a mean value, it does not reflect income distribution. If the distribution of income within a country is skewed, a small wealthy class can increase per capita income far above that of the majority of the population. In this respect Median income is a more useful measure of prosperity than per capita income, because it is less influenced by the outliers.
- Economic activity that does not result in monetary income, such as services provided within the family, or for barter, is usually not counted. The importance of these services varies widely among different economies.

## See also

- List of countries by GDP (nominal) per capita—GDP at market or government official exchange rates per inhabitant
- List of countries by GDP (PPP) per capita—GDP calculated at purchasing power parity (PPP) exchange per inhabitant
- List of countries by GNI (nominal, Atlas method) per capita
- List of countries by GNI (PPP) per capita
- List of countries by income equality
- Total personal income

# Poverty_threshold

The **poverty threshold**, or **poverty line**, is the minimum level of income deemed adequate in a given country.[1] In practice, like the definition of poverty, the official or common understanding of the poverty line is significantly higher in developed countries than in developing countries.[2] [3] The common **international poverty line** has in the past been roughly $1 a day.[4] In 2008, the World Bank came out with a revised figure of $1.25 at 2005 purchasing-power parity (PPP).[5]

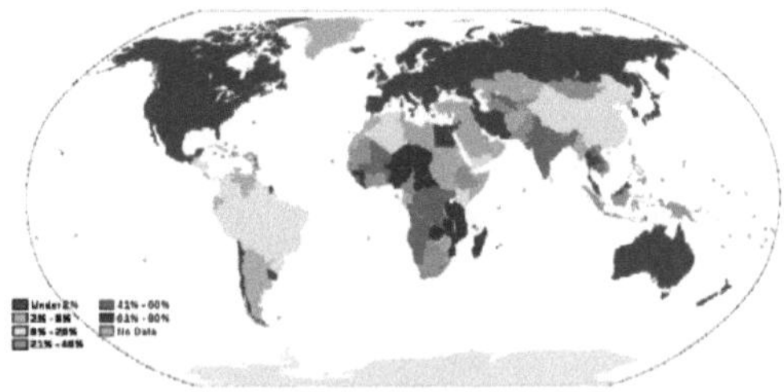

Percentage of population living on less than $1.25 per day. UN estimates 2000-2007.

Determining the poverty line is usually done by finding the total cost of all the essential resources that an average human adult consumes in one year.[6] The largest of these expenses is typically the rent required to live in an apartment, so historically, economists have paid particular attention to the real estate market and housing prices as a strong poverty line affector. Individual factors are often used to account for various circumstances, such as whether one is a parent, elderly, a child, married, etc. The poverty threshold may be adjusted annually.

## History

Charles Booth, a pioneering investigator of poverty in London at the turn of the 20th century, popularised the idea of a *poverty line*, a concept originally conceived by the London School Board.[7] Booth set the line at 10 to 20 shillings, which he considered to be the minimum amount necessary for a family of 4 or 5 people to subsist.[8] Benjamin Seebohm Rowntree (1871–1954), a British sociological researcher, social reformer and industrialist, surveyed rich families in York and drew a poverty line in terms of a minimum weekly sum of money "necessary to enable families ... to secure the necessaries of a healthy life", which included fuel and light, rent, food, clothing, and household and personal items. Based on data from leading nutritionists of the period he calculated the cheapest price for the minimum calorific intake and nutritional balance necessary before people got ill or lost weight. He considered this amount to set his poverty line and concluded that 27.84 percent of the total population of York lived below this poverty line.[9] This result corresponded with that from Charles Booth's study of poverty in London and so challenged the view, commonly held at the time, that abject poverty was a problem particular to London and was not widespread in the rest of Britain. Rowntree distinguished between primary poverty, those who lacking in income and secondary poverty, those who had enough income but spent it elsewhere (1901:295-6).[9]

## Absolute poverty

Absolute poverty is a level of poverty as defined in terms of the minimal requirements necessary to afford minimal standards of food, clothing, health care and shelter.[10] For the measure to be absolute, the line must be the same in different countries, cultures, and technological levels. Such an absolute measure should look only at the individual's power to consume and it should be independent of any changes in income distribution. The intuition behind an absolute measure is that mere survival takes essentially the same amount of resources across the world and that everybody should be subject to the same standards if meaningful comparisons of policies and progress are to be made. Notice that if everyone's real income in an economy increases, and the income distribution does not change, absolute poverty will decline.

Measuring poverty by an absolute threshold has the advantage of applying the same standard across different locations and time periods: it makes comparisons easier. On the other hand, it suffers from the disadvantage that any absolute poverty threshold is to some extent arbitrary; the amount of wealth required for survival is not the same in all places and time periods. For example, a person living in far northern Scandinavia requires a source of heat during colder months, while a person living on a tropical island does not.

This type of measure is often contrasted with measures of relative poverty (see below), which classify individuals or families as "poor" not by comparing them to a fixed cutoff point, but by comparing them to others in the population under study.

The term *absolute poverty* is also sometimes used as a synonym for extreme poverty. Absolute poverty is the absence of enough resources (such as money) to secure basic life necessities.

According to a UN declaration that resulted from the World Summit on Social Development in Copenhagen in 1995, absolute poverty is "a condition characterised by severe deprivation of basic human needs, including food, safe drinking water, sanitation facilities, health, shelter, education and information. It depends not only on income but also on access to services."[]

David Gordon's paper, "Indicators of Poverty & Hunger", for the United Nations, further defines absolute poverty as the absence of any two of the following eight basic needs:[]

- *Food:* Body Mass Index must be above 16.
- *Safe drinking water:* Water must not come solely from rivers and ponds, and must be available nearby (less than 15 minutes' walk each way).
- *Sanitation facilities:* Toilets or latrines must be accessible in or near the home.
- *Health:* Treatment must be received for serious illnesses and pregnancy.
- *Shelter:* Homes must have fewer than four people living in each room. Floors must not be made of dirt, mud, or clay.
- *Education:* Everyone must attend school or otherwise learn to read.
- *Information:* Everyone must have access to newspapers, radios, televisions, computers, or telephones at home.
- *Access to services:* This item is undefined by Gordon, but normally is used to indicate the complete panoply of education, health, legal, social, and financial (credit) services.

For example, a person who lives in a home with a mud floor is considered severely deprived of shelter. A person who never attended school and cannot read is considered severely deprived of education. A person who has no newspaper, radio, television, or telephone is considered severely deprived of information. All people who meet any two of these conditions — for example, they live in homes with mud floors *and* cannot read — are considered to be living in absolute poverty.

## Relative poverty

Relative poverty is the "most useful measure for ascertaining poverty rates in wealthy developed nations".[] [11] [] [12] [] Relative poverty measure is used by the United Nations Development Program (UNDP), the United Nations Children's Fund (UNICEF), the Organisation for Economic Co-operation and Development (OECD) and Canadian poverty researchers.[] [11] [] [12] [] In the European Union, the "relative poverty measure is the most prominent and most–quoted of the EU social inclusion indicators."[]

"[R]elative poverty reflects better the cost of social inclusion and equality of opportunity in a specific time and space."[13]

"Once economic development has progressed beyond a certain minimum level, the rub of the poverty problem – from the point of view of both the poor individual and of the societies in which they live – is not so much the effects of poverty in any absolute form but the effects of the contrast, daily perceived, between the lives of the poor and the lives of those around them. For practical purposes, the problem of poverty in the industrialized nations today is a problem of relative poverty (page 9)."[14] [13]

## History of the concept of relative poverty

In 1776 Adam Smith in the Wealth of Nations argued that poverty is the inability to afford, "not only the commodities which are indispensably necessary for the support of life, but whatever the custom of the country renders it indecent for creditable people, even of the lowest order, to be without."[] [15]

In 1958 J. K. Galbraith argued that, "People are poverty stricken when their income, even if adequate for survival, falls markedly behind that of their community."[] [15]

In 1964 in a joint committee economic President's report in the United States, Republicans endorsed the concept of relative poverty. "No objective definition of poverty exists.... The definition varies from place to place and time to time. In America as our standard of living rises, so does our idea of what is substandard."[16] [15]

In 1965 Rose Friedman argued for the use of relative poverty claiming that the definition of poverty changes with general living standards. Those labelled as poor in 1995, would have had "a higher standard of living than many labelled not poor" in 1965.[17] [15]

In 1979, British sociologist, Peter Townsend published his famous definition, "individuals [...] can be said to be in poverty when they lack the resources to obtain the types of diet, participate in the activities and have the living conditions and amenities which are customary, or are at least widely encouraged or approved, in the societies to which they belong (page 31)."[]

Brian Nolan and Christopher T. Whelan of the Economic and Social Research Institute (ESRI) in Ireland explained that, "[P]overty has to be seen in terms of the standard of living of the society in question."[]

Relative poverty measures are used as official poverty rates by the European Union, UNICEF and the OEDC. The main poverty line used in the OECD and the European Union is based on "economic distance", a level of income set at 60% of the median household income.[18]

## Relative poverty versus inequality

A measure of *relative poverty* defines "poverty" as being below some relative poverty threshold. For example, the statement that "those individuals who are employed and whose household equivalised disposable income is below 60% of national median equivalised income are poor" uses a relative measure to define poverty.[] Using this definition, if everyone's real income in an economy increases, but the income distribution stays the same, then the rate of relative poverty will also stay the same.

This means, by its very nature, that there will always be a family living in (relative) poverty, even if they have a very high living standard, unless everyone has almost exactly the same income.

Relative poverty measurements can sometimes produce odd results, especially in small populations. For example, if the median household in a wealthy neighborhood earns US$1 million each year, then a family that earns US$100,000 would be considered poor on the relative poverty scale, even though such a family could meet all of its basic needs and much more. At the other end of the scale, if the median household in a very poor neighborhood earned only 50% of what it needs to buy food, then a person who earned the median income would not be considered poor on a relative poverty scale, even though the person is clearly poor on an absolute poverty scale.

Measures of relative poverty are almost the same as measuring income inequality: If a society gets a more equal income distribution, relative poverty will fall. They point out that if society changed in a way that hurt high earners more than low ones, then relative poverty would decrease, but every citizen of the society would be worse off. Likewise in the reverse direction: it is possible to reduce absolute poverty while increasing relative poverty.

The term *relative poverty* can also be used in a different sense to mean "moderate poverty" – for example, a standard of living or level of income that is high enough to satisfy basic needs (like water, food, clothing, shelter, and basic health care), but still significantly lower than that of the majority of the population under consideration[19] .

## Basic needs

See article Basic needs

The basic needs approach is one of the major approaches to the measurement of absolute poverty in developing countries. It attempts to define the absolute minimum resources necessary for long-term physical well-being, usually in terms of consumption goods. The poverty line is then defined as the amount of income required to satisfy those needs. The 'basic needs' approach was introduced by the International Labour Organization's World Employment Conference in 1976.[] [] "Perhaps the high point of the WEP was the World Employment Conference of 1976, which proposed the satisfaction of basic human needs as the overriding objective of national and international development policy. The basic needs approach to development was endorsed by governments and workers' and employers' organizations from all over the world. It influenced the programmes and policies of major multilateral and bilateral development agencies, and was the precursor to the human development approach."[] []

A traditional list of immediate "basic needs" is food (including water), shelter and clothing.[] Many modern lists emphasize the minimum level of consumption of 'basic needs' of not just food, water, and shelter, but also sanitation, education, and health care. Different agencies use different lists.

In 1978 Ghai investigated the literature that criticized the basic needs approach. Critics argued that the basic needs approach lacked scientific rigour; it was consumption-oriented and anti-growth. Some considered it to be "a recipe for perpetuating economic backwardness" and for giving the impression "that poverty elimination is all too easy".[] Amartya Sen focused on 'capabilities' rather than consumption.

In the development discourse, the basic needs model focuses on the measurement of what is believed to be an eradicable level of poverty.

Professor Chris Sarlo, an economist at Nipissing University in North Bay, Canada, and a senior fellow of the Fraser Institute, used Statistics Canada's socio-economic databases, particularly the *Survey of Household Spending* to determine the cost of a list of household necessities. The list includes food, shelter, clothing, health care, personal care, essential furnishings, transportation and communication, laundry, home insurance, and miscellaneous; it assumes that education is provided freely to all residents of Canada. This is calculated for various communities across Canada and adjusted for family size. With this information, he determines the proportion of Canadian households that have insufficient income to afford those necessities. Sarlo introduced his basic needs approach in 1992 in his book entitled *Poverty in Canada* published by the Fraser Institute. Sarlos argued that Statistics Canada's Low Income Cutoff (LICO), which uses a relative poverty line threshold was inadequate. Sarlos' basic needs poverty measure for Canada was updated in 2001 Sarlos' to reflect "new developments in the literature, external critiques of the approach, and better (and more comprehensive) data". For example it included out-of-pocket health-care costs. According to the Sarlo, "the basic-needs approach is partly absolute (the list [of necessities] is limited to items required for long-term physical well-being) and partly relative, reflecting the standards that apply in the individual's own society at the present time." It was updated again in 2006. [] The Fraser Institute's list of necessities for living creditably in Canada includes not only food, shelter, clothing, and health care, but also personal care, furniture, transportation, communication, laundry, and home insurance. It is criticized for not including any entertainment items like cable television, daily newspapers, and tickets to movies or sporting events.[] [] [20] [20] Based on his basic needs poverty threshold, the poverty rate in Canada, the poverty rate has declined from about 12% of Canadian households to about 5% since the 1970s.[21] This is in sharp contrast to Statistic Canada, the Conference Board of Canada, the Organisation for Economic Co-operation and Development (OECD) and UNESCO reports on poverty in Canada. They used the relative poverty measure considered to the most useful for advanced industrial nations like Canada, which Sarlo rejects.

According to a 2008 report by the Organisation for Economic Co-operation and Development (OECD), the rate of poverty in Canada, is among the highest of the OECD member nations, the world's wealthiest industrialized nations.[] There is no official government definition and therefore, measure, for poverty in Canada. However, Raphael Dennis, author of *Poverty in Canada: Implications for Health and Quality of Life*[] [] reported that the

United Nations Development Program (UNDP), the United Nations Children's Fund (UNICEF), the Organisation for Economic Co-operation and Development (OECD) and Canadian poverty researchers[22] [] find that relative poverty is the "most useful measure for ascertaining poverty rates in wealthy developed nations such as Canada".[] [11] [] [12] []

## National poverty lines

National estimates are based on population-weighted subgroup estimates from household surveys. Definitions of the poverty line may vary considerably among nations. For example, rich nations generally employ more generous standards of poverty than poor nations. Thus, the numbers are not comparable among countries.

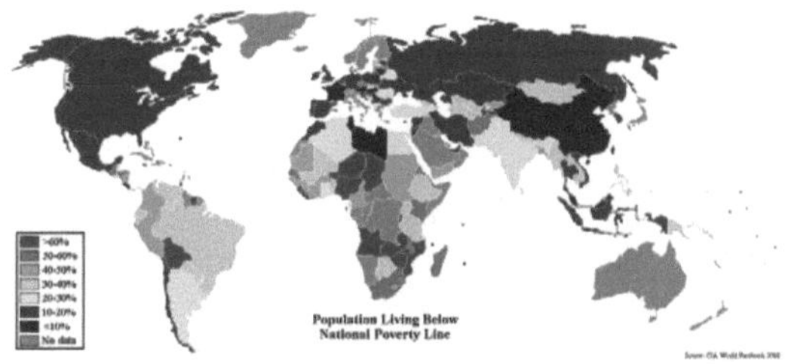


CIA World Factbook-based map showing the percentage of population by country living below that country's official poverty line.

In 2010, in the United States, the poverty threshold for one person under 65 was US$11,344 (annual income); the threshold for a family group of four, including two children, was US$22,133.[] [23] According to the U.S. Census Bureau data released on September 13, 2011, the nation's poverty rate rose to 15.1 percent in 2010.

In the UK, "more than five million people – over a fifth (23 percent) of all employees – were paid less than £6.67 an hour in April 2006. This is based on a low pay rate of 60 percent of full-time median earnings, equivalent to a little over £12,000 a year for a 35-hour working week. In April 2006, a 35-hour week would have earned someone £9,191 a year – before tax or National Insurance".[24] [25]

India's official poverty level, on the other hand, is split according to rural vs. urban thresholds. For urban dwellers, the poverty line is defined as living on less than 538.60 rupees (approximately USD $12) per month, whereas for rural dwellers, it is defined as living on less than 356.35 rupees per month (approximately USD $7.50).[]

## Criticisms

Using a poverty threshold is problematic because having an income marginally above it is not substantially different from having an income marginally below it: the negative effects of poverty tend to be continuous rather than discrete, and the same low income affects different people in different ways. To overcome this problem, poverty indices are sometimes used instead; see income inequality metrics.

A poverty threshold relies on a quantitative, or purely numbers-based, measure of income. If other human development-indicators like health and education are used, they must be quantified, which is not a simple (if even achievable) task.

### Overstating poverty

In-kind transfers, whether from public or private sources, are not counted when calculating a poverty threshold. For example, if a parent pays the rent on an apartment for an adult son directly to the apartment owner, instead of giving the money to the son to pay the rent, then that money does not count as income to the son. If a church or non-profit organization gives food to an elderly person, the value of the food is not counted as income to the elderly person. Rea Hederman, a senior policy analyst in the Center for Data Analysis at the Heritage Foundation, said

> The official poverty measure counts only monetary income. It considers antipoverty programs such as food stamps, housing assistance, the Earned Income Tax Credit, Medicaid and school lunches, among

others, "in-kind benefits" – and hence not income. So, despite everything these programs do to relieve poverty, they aren't counted as income when Washington measures the *poverty rate.*[26]

Studies measuring the difference between income before and after taxes and government transfers, however, have found that without these programs poverty would be roughly 30% to 40% higher than the official poverty line indicates, despite many of their benefits not being counted as income.[27] [28]

Further, the U.S. Census Bureau calculates the poverty line the same throughout the U.S. regardless of the cost-of-living in a state or urban area. For instance, the cost-of-living in California, the most populous state, was 1.42 times the U.S. average in 2010 while the cost-of-living in Texas, the second most populous state, was 0.90 (90%) of the U.S. average.

## Notes

[1] Ravallion, Martin *Poverty freak: A Guide to Concepts and Methods*. Living Standards Measurement Papers, The World Bank, 1992, p. 25

[2] Hagenaars, Aldi & de Vos, Klaas *The Definition and Measurement of Poverty*. Journal of Human Resources, 1988

[3] Hagenaars, Aldi & van Praag, Bernard *A Synthesis of Poverty Line Definitions*. Review of Income and Wealth, 1985

[4] Sachs, Jeffrey D. *The End of Poverty* 2005, p. 20

[5] Ravallion, Martin; Chen Shaohua & Sangraula, Prem *Dollar a day* The World Bank Economic Review, 23, 2, 2009, pp. 163-184

[6] {Poverty Lines-Martin Ravallion, in The New Palgrave Dictionary of Economics, 2nd Edition, London: Palgrave Macmillan}

[7] Alan Gillie, 'The Origin of the Poverty Line', Economic History Review, XLIX/4 (1996), 726

[8] David Boyle - The Tyranny of Numbers p.116

[9] page 298

[11] Child poverty in rich nations: Report card no. 6 (Report). Innocenti Research Centre. 2005.

[12] Human development report: Capacity development: Empowering people and institutions (Report). Geneva: United Nations Development Program. 2008.

[13] Jonathan Bradshaw; Yekaterina Chzhen; Gill Main; Bruno Martorano; Leonardo Menchini; Chris de Neubourg (January 2012) (PDF). [www.unicef-irc.org/publications/pdf/iwp_2012_01.pdf Relative Income Poverty among Children in Rich Countries] (Report). Innocenti Working Paper. Florence, Italy: UNICEF Innocenti Research Centre. ISSN 1014-7837. www.unicef-irc.org/publications/pdf/iwp_2012_01.pdf.

[14] A League Table of Child Poverty in Rich Nations (Report). Innocenti Report Card No.1. Florence, Italy: UNICEF Innocenti Research Centre.

[15] Peter Adamson (2012) (PDF). Measuring child poverty: New league tables of child poverty in the world's rich countries (http://www.unicef.ca/sites/default/files/imce_uploads/DISCOVER/OUR WORK/ADVOCACY/DOMESTIC/POLICY ADVOCACY/DOCS/unicefreportcard10-eng.pdf) (Report). UNICEF Innocenti Research Centre Report Card. Florence, Italy. .

[16] Minority [Republican] views, p. 46 in U.S. Congress, Report of the Joint Economic Committee on the January 1964 Economic Report of the President with Minority and Additional Views (Report). Washington, D.C.: US Government Printing Office. January 1964.

[17] Friedman, Rose. D. (1965). " *Poverty: Definition and Perspective* ". *American Enterprise Institute for Public Policy Research* (Washington, D.C.).

[19] http://www.caledonia.org.uk/papers/Who-and-Where-are-the-Poor.pdf

[20] Chris Sarlo (23 July 2001). Measuring Poverty in Canada (Report). Fraser Institute.

[21] Poverty in Canada: 2006 Update (http://www.fraserinstitute.org/COMMERCE.WEB/product_files/PovertyinCanada2006.pdf)

[22] The Conference Board of Canada "uses the OECD's relative measure of child poverty, which calculates the proportion of children living in households where disposable income is less than 50 per cent of the median in each country." The Conference Board 2013 cautioned that Canada's high poverty rate, ranks among the worst of the 17 countries they compared. "Canada's child poverty rate was 15.1 per cent, up from 12.8 per cent in the mid-1990s. Only the United States ranked lower.

[24] Working out of Poverty: A study of the low paid and the working poor by Graeme Cooke and Kayte Lawton

[25] IPPR Article: "Government must rescue 'forgotten million children' in poverty" (http://www.ippr.org.uk/pressreleases/?id=2965)

[26] "Poor poverty yardsticks" (http://www.heritage.org/Press/Commentary/ed090806b.cfm), Rea S. Hederman, Jr., September 8, 2006, The Heritage Foundation

[27] Kenworthy, L. (1999). Do social-welfare policies reduce poverty? A cross-national assessment. *Social Forces, 77*(3), 1119-1139. doi:10.1093/sf/77.3.1119

[28] Bradley, D., Huber, E., Moller, S., Nielson, F. & Stephens, J. D. (2003). Determinants of relative poverty in advanced capitalist democracies. (http://www.jstor.org/stable/3088901) *American Sociological Review, 68*(3), 22-51.

## See also

- Asset poverty
- Income deficit
- List of countries by percentage of population living in poverty
- Living wage
- Minimum wage
- Measuring poverty
- UN Millennium Development Goals

## References

## Further reading


- Alan Gillie, "The Origin of the Poverty Line", *Economic History Review,* XLIX/4 (1996), 726


## External links

- History of the U.S. Poverty Line (http://eesc.orst.edu/agcomwebfile/edmat/html/em/em8743/part1/officialgovt.html) by Tom Gentle, Oregon State University.
- United States Department of Health and Human Services Poverty Guidelines, Research, and Measurement (http://aspe.hhs.gov/poverty/index.shtml)
- 2007 United States Department of Health and Human Services Poverty Guidelines (http://aspe.hhs.gov/poverty/07poverty.shtml)
- Debraj Ray 1998, *Development Economics*, Princeton University Press, ISBN 0-691-01706-9.
- World Summit for Social Development Agreements (http://www.un.org/esa/socdev/wssd/agreements/index.html), United Nations
- The "elimination" of poverty (http://www.inclusivedemocracy.org/journal/vol4/vol4_no1_takis_poverty.htm), Takis Fotopoulos, *The International Journal of Inclusive Democracy* (http://www.inclusivedemocracy.org/journal/), vol.4, no.1 (January 2008).

# Article Sources and Contributors

**Richland,_Pennsylvania** *Source*: http://en.wikipedia.org/w/index.php?title=Richland%2C_Pennsylvania *Contributors*: 2001:558:6031:1D:411:4F86:7E24:8D93, Gerry D, Grandwizard, Hmains, Jknoll52, Jllm06, John Cardinal, Nyttend, PurpleChez, Ram-Man, Robertjohnsonrj, VerruckteDan, 7 anonymous edits

**Borough** *Source*: http://en.wikipedia.org/w/index.php?title=Borough *Contributors*: 130.225.29.xxx, A8UDI, AFBorchert, Acerperi, Adam Carr, After Midnight, Ahkitj, AjaxSmack, Alansohn, Alhen, Altenmann, Angr, Anjwalker, Anthony Appleyard, Antipastarasta, Anywikiuser, Aranel, ArielGold, Aripap, Astronaut, Atrian, Augustusr, BOARshevik, BRG, Bcorr, Bearcat, Beland, Belfry, Ben Arnold, Bermicourt, Biopresto, Bkonrad, BlueAg09, Bob Burkhardt, Branddobbe, Brandon5485, Brian1975, Brozen, CORNELIUSSEON, Canaima, Canking, Canterbury Tail, Cassowary, Cecropia, Chameleon, Chmod007, Chowbok, ChrisGualtieri, Chrislk02, Christian Taylor, Cjmnyc, ColinFine, Colonies Chris, Computor, Comrade42, Conversion script, Cornischong, CrimsonBlue, Cryptic, Dac04, Dale Arnett, Danrok, Dave souza, David Kernow, Deflective, Derek Ross, Djflem, Doctor Whom, Dolovis, Dpaajones, Dramatic, Drphilharmonic, Dubidub, E-pen, Editorxd, Eep², Eja2k, Erianna, Eric, Esrever, Euchiasmus, Exile, Fayenatic london, Fortdj33, FreplySpang, FutureNJGov, G-Man, Galoubet, Gene.arboit, Giraffedata, Glover, Grafen, Graham87, GrammarFascist, Greba, Ground Zero, Guilloguzman, Hede2000, Hedgehog, Heegoop, Henrygb, Herr Ratselhaft, Hibernicus, Hneto, Howcheng, Izzycat, JBC3, James Crippen, JanCeuleers, Java7837, Jcmenal, Jed keenan, JerryFriedman, Jimp, JoanneB, Joeldl, John, Jondel, Joseph Solis in Australia, Jules7484, Julipanette, Justinsomnia, KKramer, Kalamkaar, Kazvorpal, Keeno, Kevin McE, Kookyunii, Kwamikagami, Langleap gal, Leslie Mateus, Lexicon, Lexusuns, Livajo, Loonymonkey, Lozleader, MRSC, MSSEVER, MacTire02, Mais oui!, MapsMan, Mboverload, Menchi, Mhallmusic, Mhockey, Mike65535, MinisterForBadTimes, Mirror13, Modulatum, Monosig, Montrealais, Morwen, Mothman, Mr Stephen, Muniswede, Murphymc, NHRinger, Namiba, Nesbit, Neutrality, NickNak, Ninetyoneschool, Nricardo, Nyttend, Oakwood, OliverR, Omegatron, Owain, PBS-AWB, Pavel Vozenilek, Peak, Peter Horn, Pfly, Pmanderson, R'n'B, RJP, Ram-Man, Rbrwr, Rcingham, Redf0x, Richard New Forest, Richardguk, Robhmac, Robminchin, Rrius, RussBlau, Russianamerican1, S19991002, Saga City, Saltmarsh, Samuelsen, Sardanaphalus, Sashal, ScottAlanHill, Sensor, Shenme, Shepard, Silverhelm, Sjorford, Sochwa, Someguy1221, Staffelde, Station1, Stephen C. Carlson, StudentJCase, Tb, Tbone, TelecomNut, The Tom, The Transhumanist, ThunderingTyphoons!, Tobias Conradi, Tobias Hoevekamp, Tolly4bolly, Tosek, Travelbird, TurkChan, Ulric1313, User№101, Whytecypress, Wighson, William Allen Simpson, Winhunter, Woohookitty, Zero Gravitas, 184 anonymous edits

**Lebanon_County,_Pennsylvania** *Source*: http://en.wikipedia.org/w/index.php?title=Lebanon_County%2C_Pennsylvania *Contributors*: Acntx, Altairisfar, Anna199123, AshyLarry, Avraham, Bob Burkhardt, Boothy443, Brybry1999, CanisRufus, Caponer, CommonsDelinker, Compfreak7, Crusadinggoonie, D6, Denisarona, Derekbridges, Dewritech, Distcoach2, Dneubert, Dthomsen8, Elihu03, FF2010, Gehirn, Gerry D, Giupèt, Heff01, Hephaestos, Hyperfast, IsWayneBradygonnahavetosmackabitch, Itlnstln921, Krugerhlu, Leslie Mateus, Lightmouse, Monegasque, NERIUM, Nuge, Nyttend, Ohnoitsjamie, OldsVistaCruiser, Omnedon, Pearle, Ram-Man, Robertjohnsonrj, Ruhrfisch, Scottfisher, Search4Lancer, Sgt Pinback, Smallbones, Spork232, StAnselm, T. Anthony, TUF-KAT, Template namespace initialisation script, Tim!, Utopies, VerruckteDan, Vinegar, WilliamJE, Worldenc, Wrightchr, Z4ngetsu, 36 anonymous edits

**United_States** *Source*: http://en.wikipedia.org/w/index.php?title=United_States *Contributors*: !Silent, "Country" Bushrod Washington, (, (aeropagitica), *hollywood*, -Barry-, 041744, 067pansaol, 08OceanBeach SD, 08murtaghkc, 09LFIRTH, 0Bammer, 0alx0, 0ts0, 1.21 jigwatts, 1028, 10nn, 111Alleskönner, 119, 123Mike456Winston789, 1251thestrokes, 12835S, 1297, 149AFK, 1658, 16@r, 172, 18alex12, 19994best, 19starwars77, 1oddbins1, 1of3, 2005-03-04T23:39Z, 2005-03-09T21:41Z, 2005-04-08T13:18Z, 212etc, 23 editor, 23haveblue, 23prootie, 25 Cents FC, 2812, 28421u2232nfenfcenc, 2nd Piston Honda, 2t56, 3 years too many, 3232330, 400, 50imp, 5ptcalvinist, 621PWC, 66.81.31.xxx, 6r6r6muyr6uh, 7i5, 7mike5000, 84user, 8r1ck, 90, 987li, A Clown in the Dark, A Softer Answer, A State Of Trance, A Train, A bit iffy, A s williams, A-b-c-d-e-f, A-giau, A520, AAA765, ACD605, ACSE, ACW, ACupOfCoffee, AED, AEMoreira042281, AIRBUS, AKGhetto, AKMask, ALEXF971, AP1787, ARC Gritt, ARCoolguy, ARUenergy, ASDFGH, ASOTMKX, AStudent, AVM, AZCactus1, Aaker, Aalahazrat, Aang-kai, Aaporter 87, Aaron DT, Aaron Einstein, Aaron Schulz, Aaron mcd, Aaron7chicago, AaronS, AaronY, Abayarde, Abc is a number, Abce2, Abdalla A, AbeOlsen, Abecedare, Abilityfun, Abog, Abonazzi, Abtract, Abuchman1, Academic Challenger, Acalamari, Acanz, Accepttruth, Accurizer, Ace ETP, Acerperi, Acetic Acid, Acio83, Acjelen, Acmilan15, Acroterion, Adam Bishop, Adam Pamilih, Adam1213, AdamFirst, AdamJacobMuller, AdamRetchless, Adamski24, Adamsmo, Adamstraw99, Adanedhel21, Adashiel, Adavidb, Addicted04, Addisonn., Addit, Adelina and Hannah, Adhall, AdjustShift, Adjwilley, Adko110, Adnan bogi, Adolphus79, Adoniscik, Adpete, Adraeus, Adriaan, Adrian, Adrimaster3, AdultSwim, Aenar, Aeonx, Aerol, Aesopos, Aeyl13, Afil, Afinebalance, Afropuffs71, Against the current, Agdaman4life, Agtaz, Aherunar, Ahlabonde, Ahmad.ibn.as.Sayyid, Ahmadyk23, Ahodes1, Ahoerstemeier, Ahuskay, Aidepolcycne, Aido2002, Aim Here, Aiman abmajid, Aimjeff2, Airdish, Aislar, Aivazovsky, AjaxSmack, Ajwitte, Akamad, Akanemoto, AkvinRox, Akyoyo94, Al Villarruel, Al-Andalus, Al1encas1no, Alakasam, Alan Flynn, Alan J. Villarruel, Alan W, Alan.ca, AlanBarrett, AlanM1, Alandavidson, Alanhaggai, Alanscottwalker, Alansohn, Alarics, Albert0057, AlbertR, Albinostorm, Alcuin, Aldie, AlefZet, Alegoo92, Alephcat, Alex S, Alex43223, AlexJP, AlexNebraska, AlexOUK, Alexander Vince, Alexdragon, Alexignatiou, Alexihlo, Alexnovo, Alexsautographs, Alexwcovington, Alfiboy, Alhutch, Ali Karbassi, Ali a, Ali944rana, Aliazimi, Alientraveller, Alienus, Alilhappything, Alishakitty, Aliwaseem, Alkivar, All Male Action, All in, AlltimeJourneyFan, AllyUnion, AlmasterGM, Alodyne, AlphaEta, Alphabetagamnma, Altanar, Altenmann, AltiusBimm, Alvin Lee, Alvinrune, Amberrock, Amcfreely, AmericaIsNumberOne, AmericanXplorer13, Americanization, AmiDaniel, Amikeco, Aminatam, Amirb1, Amp873, Amren, An Siarach, AnOddName, Anakus, Anaraug, Anastrophe, Anativecantonesespeaker, Ancheta Wis, Andersmusician, Andkore, Andonic, Andre Engels, Andrea moro, Andrejj, AndresTM, Andrevan, Andrew Gray, Andrew Levine, Andrew Steller, Andrew c, Andrewbarber79, Andrewman327, Andreworkney, Andrewphelps, Andrewpmk, Andrewsthistle, Andris, Android79, Andy Marchbanks, Andy-ch, Andy120290, AndyBQ, AndyZ, Andyabides, Andyb215, Andychen86, Andypandy.UK, Andypham3000, Andyso, Anetheron589, Angel David, Angela, Angelastic, Angelboy, Angelo De La Paz, Angr, AngryParsley, Angryhobo16, Angusmclellan, Anhydrobiosis, Anilocra, Animum, Aninnymouse, Anirvan, Ankhet, Ankit jn, Ankitsingh83, Ann O'nyme, Anna Lincoln, Annama29, Annandale, Annasalia, AnnicAllus, Anon6666, AnonMoos, AnonyLog, Anonymity001, Anonymous editor, AnselmoLacerda, Anshumsaxena, Antandrus, Antarticstargate, Anthere, Anthony, Anthony Appleyard, Anthony717, Anti-Wuss, AntiBodies, AntiBones, AntiVanMan, Antimatter15, Anto475, Anton27, AntonioB, Antopi, Anubis2051, Anujill, Anywikiuser, Anzoooo, Aogouguo, Ap, Apatens, ApolloBoy, Apostrophe, Appleboy, Applemaster, ApprenticeFan, Aptdwn26, Apwvt, AquaD, Arad, Araiderplaya, Aranel, Aravind V R, Arbor to SJ, Arbuge, Arcadie, Arch dude, Archaelicos, Archaic, Archimedez, ArchonMeld, Arcimpulse, Arctic.gnome, AreJay, AreaControl, Argentsky, ArglebargleIV, AriGold, Aricci526, Aris Katsaris, Aristophanes68, Aristotle1990, Arkalochori, Armenia81, Armyrifle9, Arnon Chaffin, Arnoutf, Arpingstone, Arsenio, Arshadm, Art LaPella, Arthur Chevailier, Artifactblue, Artimbo, Artystyk386, Arwack, Arwel Parry, Arx Fortis, Aryavib, Asbestos, AscendedAnathema, Asderoff, Asdfjkl123, Asenai, AshAloy, Ashenai, Ashershow1, Ashnard, Ashrichards90, Ashton 29, Aspensti, AspiringScholar, Assyria 90, Asterix70, Astolfo Appleyard, Astonvilla91, AstroNomer, Astronautics, Astros4477, Astrotrain, Atchom, Atemperman, Athenean, AtilimGunesBaydin, Atlant, Atmans, Atomique, Atoric, AuburnPilot, Auburnirishman, Audacity, Aude, Aufheben, AuroraReloaded, Auroranorth, Ausimus, Austen Redman, Austen rules, Austin Hair, AustinKnight, Auswiger, Author782, AutomaticStrikeout, Autonova, Avala, AvalonXQ, Avargasm, Avaya1, Avenue, Avg, Avicennasis, Avidius, Avsrule1921, Awilcox1533, AxSkov, Axel Svensson, Axeman89, Aximill, Aydchery, Ayla, Ayrton Prost, AzaToth, Azmurath, B-watchmework, B. Fairbairn, B0at, B0mb3d, B1link82, B64, BD2412, BECASC, BG7, BGManofID, BGordon, BJCenters, BMF81, BRG, BSveen, Baa, Babablacksheephaveyouanywool, BabbaQ, Babbage, Babbby, Babygrand1, Babylicious2, Backslash Forwardslash, Backwalker, Badanagram, Badbe0, Bahn1k, Balderdash707, Ballboys10, Ballestrain, Bambuway, Bamford, Banders234, Bandity, Banes, Bansp, BanyanTree, Baplexed, Barbara Shack, Barbieblader, Barek, Barfooz, BaronLarf, Barrylb, Barryob, Bart133, BarzaniKurd16, Basbalfrk, Baseball Bugs, Baseballbaker23, Bash RF, Basilisk4u, Basohead, Bassbasketball03, Bassplr19, Bastique, Battem, Baxter 10, Bayonetblaha, Bazonka, Bbatsell, Bbenjoe, Bbert1994, Bbert1994cchs, Bbx, Bchalfin, Bdesham, Beach bum7, Beamathan, Beamer103, Bearingafish, Beastcuber, Bedford, Beef cakes, Beefy247, Beepoppab, Beetle1042000@yahoo.com, Beezer137, Before My Ken, Beh-nam, Behemoth98, Beland, Belchfire, Belligero, Beloki, Ben Arnold, Ben D., Ben Lunsford, Ben76266, BenFrantzDale, BenH, Benbristol, Benc, Benhocking, BenjaminTsai, Benji Powell, Benjiboi, Benkalt, Benlisquare, Benson85, Bentobias, Benwbrum, Berek Halfhand, Bet1973, Beta m, Betmeno, Bevo, Bevo74, Bgwhite, Bhadani, BhaiSaab, Bhamv, Bhe22, Bibbledebobblede, Bibbly Bob, Bidgee, Big Adamsky, BigBoyRubio, BigFishy, Bigairboarder121, Bigbear bh, Bigbluefish, Bigferdes, BiggKwell, Bigjust12345, Bigmak, Bigmanly, Bigredbrain, Bihco, Bilby, Bility, Bill McCloskey, Bill william compton, Bill-on-the-Hill, BillC, Billheller, Billim1, Billrnyc, BillyBoom, Billymac00, Binkhaldun, Biohazard930, Bipalabras, Bird1455, Birdmessenger, Birdofprey, Birthofthehive, Biruitorul, Bishonen, Bisteca, BitWarrior, Bjarki S, Bjmarfito, Bjoh249, Bjpremore, Bkell, Bkkeim2000, Bkonrad, Bky1701, Black Kite, Black and White, Black lyzard, BlackHades, Blackanthrope, Blackcell16, Blackmongoose, Blackshafen, Blainster, BlankVerse, Blarg2.0, Blastwizard, Blaylockjam10, Bleff, Blessthishouse, Blethering Scot, Blindman shady, BlisteringFreakachu, Blog Mav Rick, Blondace, Blood-Angel 9th Ranger, Blu Aardvark, Blue7545, BlueAg09, BlueSalo, Bluedog423, Bluedogtn, Bluegold, Bluemask, Bluemoose, Bluwiki123, BlytheG, Bmdavll, Bmgoau, Bmicomp, Bmitchelf, Bo, Bob A, Bob rulz, Bob schwartz, Bob swatsky, Bob121, Bobblehead, Bobblewik, Bobbob, Bobbum 47, Bobby Mouth of the South, BobbyAFC, BobbyJ, Bobbyi, Bobchalk, Bobdobbs1723, Bobdylan3589, Bobfreshwater, Bobisbob, Bobmack89x, Bobo The Ninja, Bobo192, Bobobobo, Bobozz, Bobtheuglyhobo, Bodnotbod, Bogdan, Bohemian tosspot, Bohuiginn, Boman1990, Bondegezou, Bonewah, Boneyard90, Bongwarrior, Bonked116, Bono983, BonsaiViking, Boogaryou, Boomshadow, Boothy443, Boraczek, Bornintheguz, Bosonic dressing, BostonRedSox, Bowei Huang, Bowei Huang 2, Boxbrown, BozMo, Bporter28, Br'er Rabbit, BradM, Bradifer, Bradwart, Braffit, Brainboy109, Brainhell, Brainstorm284, Branddobbe, Brandmeister, Brandmeister (old), Brandon, Brant4, Brat32, Bratsche, Bravada, Bread for the ducks, Breisblonde22, Brendan O'Shea, Brendanconway, BrendelSignature, Brennan626, Brenton, Brian Pearson, Brian the Editor, Brian0918, Brianga, Brickfield, Brickmack, Bridg1, Brighterorange, Brilliburger, Brion VIBBER, Brisvegas, Britannic1, Britcom, British legend, BritishWatcher, BritneySpearsAddict, Britsrule06, Broadside Perceptor, Brockert, Brodey, Broken Segue, BrokenSegue, Brokenfuse, Bronx Discount Liquor, Brossow, BrotherFlounder, Bruguiea, Brutaldeluxe, Brutannica, Bryan Derksen, BryanG, Bryant693, Brycehughes, Bssc81, Btipling, Buaidh, Buddha24, Buffalob07, Bugs5382, Bulba2036, Bull Market, Bulldog123, Bulldog73, Bullshark44, Bully25, Bunny-chan, Burningview, Burntsauce, Burroughsks88, Burto88, Bustoff, Butefraim22, Butros, Butwhatdoiknow, Buzity, Buzz-tardis, Bwinn, Bwithh, C. M. Harris, C.Fred, C.J. Griffin, C.Löser, CALR, CAMorris, CASportsFan, CDHgrün, CDN99, CJ, CJ Aikman, CJ King, CJK, CJLL Wright, CKBrown1000, CLW, CMSimons, CN3777, COF, COMPFUNK2, CORNELIUSSEON, CPacker, CQJ, CRouleau, CTF83!, Cacophony, Cactus.man, Cadiomals, Cafemusique, Cafzal, CalJW, Calcprgmer10011, CalicoCatLover, Califunya, Calliopejen1, Callmemrshowoff, Calltech, Calmypal, Calu2000, Calvin Lourdes He, Cam24, Cam275, Cambrasa, CambridgeBayWeather, Camenzind, Cameron, Cameron12345, Cameron12355, Camr, Camw, Can man138, Can't sleep, clown will eat me, Canada Jack, Canadaisawesome, Canadia, CanadianCaesar, CanadianLinuxUser, Canderson7, Caniago, CanisRufus, Canofmeat, Canterbury Tail, Cantus, Canuckian89, CapitalR, CapitalSasha, Capmo, CaptainNate, Captaincoach92, Carbonite, Cardiff123098, Cardinalwraith, CaribDigita, CarlosMarti123, Carnogaunt, CarolinianJeff, Carowinds, Carptrash, Carrp, Carthradge, Casper Gutman, Castletower, Castncoot, Catdude, Catswilltakeovertheworld1day, Causa sui, Cause of death, Caveman 07, Cbrett, Cbrown285, Cburnett, Ccacsmss, Cchow2, Ccs4ever, Cdc, Cdcon, Ce garcon, Cea801, Cecil722, Cecilgol, Cedric88, Cedrus-Libani, Celestianpower, Cello0wnsy0u, Cellurl, Celyndel, Censusdata, CentralTiger74, Centrx, Cenwin88lee, Cerealkiller13, Cerebrus13, CesarB, Ceyockey, Cfkane84, Cfquinlan, Cgoodell, Ch'marr, ChKa, Chadlupkes, Chanlyn, Channard, Channelone, Chanting Fox, ChaosSorcerer91, Chaosdruid, Chaosname, Charlie chaplin, CharlotteWebb, Charruss, ChaseJ589, Chason11, Chatori, Che829, Cheatcode2, Check two you, CheckPlus, Chedorlaomer, Cheese bombs, Cheese007, Cherkash, Chessman, ChesterTheWorm, Cheung1303, Cheungpat, Chiamus, Chic-ism, Chichui, Chickenofbristol, Chicopac, Chillum, Chindia

(China-India), Chinju, Chinneeb, Chipmunk46, Chipmunkdavis, Chivista, Chobbs138, Chochem, Chochopk, Chocoforfriends, ChokeSlamFromHell, Cholmes75, Chongkian, Chris 73, Chris Mason, Chris cufc, Chris the Paleontologist, Chris-dotcom, Chris432, Chrisahn, Chriscoooley, Chrislk02, Chrismichener1, Chrisranders, Christian Historybuff, Christian List, Christinam, Christinewales, Christoph Braun, Christopher Kraus, Christopher Mulcahy, Christopher Parham, Christopher higham, Christophore, Chronus, Chrose, Chunky Rice, ChunkySoup, Chzz, CieloEstrellado, Cienade, Cieran 91, Ciroa, Cirt, Citation Needed, Citizen-of-wiki, Cityhighkid09, Civil Engineer III, Cje, Cjeiler, Ckatz, Clab6, Claidheamohmor, ClamDip, Clanget, Classic Fan, Clawson, Clayoquot, Cleared as filed, Clicketyclack, Clickpop, CliffC, Cloakdeath, Clorox, Closedmouth, CloudNine, Cloudblazer, Cloudreaver, Cma, Cmadler, Cmc0, Cmcginnis, Cmguy777, Coaster1983, Coasterlover1994, Cocacola125802, Cocoaguy, Cocoliras, Cocoma, Cod1337, Code Logic, Codingmonkey, Codu, Cofax48, Cohesion, Cold Light, ColdFusion650, ColdWarCharlie, Collect, Collier, Colonel Cow, Colonies Chris, Comayagua99, Cometstyles, Commander Keane, Commonbrick, CommonsDelinker, Compfreak7, Comput65, ComputerBuker, Computerwhiz171, Comrade Martin, Concordia, Coneslayer, Confiteordeo, Congolese fufu, Conk 9, Conman23456, Connelly, Conningcris, Connormah, Conor 33, Conquistador48, Conte di Cavour, Contemptiblepunx, Contra10, Conversion script, Conveyance, Conza1994, Cool Blue, Cool Stuff Is Cool, Cool-guy357, Cool4x411, Coolalec11, Coolcaesar, Cooldoug111, Coolhawks88, Coolio Grooverider, Cooliocheck1, Cooljuno411, Coolpeng123, Coolstorybrooo, Copperchair, Coraje, Cordless Larry, Cordova, Coredesat, Corgimaster, Corlen, Cormac Canales, Cornellrockey, Corriebertus, Corticopia, Corvun, Corwin8, Cosmic Latte, CosmologyProfessor, Costlab, Cougar Draven, Coulraphobic123, CoulterTM, Countersubject, Countmall, Courcelles, CovenantD, Coviepresb1647, Cow, Cowman109, Cpavlovski, Cprompt, Cptbuck, Cr0ssfire, Crad0010, Craigy144, Craky, Crash14, Crazy8jacky, Crazycomputers, Crazyparadigm, Crazytail2, Crazytales, Crd721, Credema, Creez34, Crevaner, Criptofcorbin, CristianoMacaluso, Critical.solvent, Crm18, CrnaGora, Crobzub, Crohall, Cronic 5085, Cros13, Crowbaaa, CrucifiedChrist, Crunchedfor6, Crusadeonilliteracy, Cruzian, Crxssi, Cryptic, CrypticBacon, CryptoDerk, Cryptosquid, Crzrussian, Cs-wolves, CsDix, Csigabi, Csurguine, Cthia, Ctryellowbird, Ctuavocado, Ctyankeeboy1638, Cuak's, CuboneKing, Cuchullain, Cuimalo, CumbiaDude, Cunado19, Curps, CurtisJohnson, Cutiewithabooty123, Cutrupi04, Cuye, Cvcdrk, Cverlo, Cyan, Cyanidethistles, Cybercavalier, Cybercobra, Cyde, Cynical, Cynicism addict, Cyp, Cypher056E, Cyril Washbrook, Cyrilthemonkey, Cyrus111, D prime, D-Rock, D.M. from Ukraine, D.illah, D54b7ba, D6, DARTH SIDIOUS 2, DBaba, DC, DCB4W, DCGeist, DEW 7, DGJM, DHN, DJ Clayworth, DJHeartAttack, DL79OL, DLJessup, DMCer, DMellis, DTSX, DVD R W, DaGizza, Dabbler, Dablaze, Dabomb87, Dabrown, Dacoslett, Dacres, Daedalus969, Dah31, Dak is back, Dale Arnett, Dale-DCX, Daledavey, Dalek, Dalstadt, Damian Yerrick, Damicatz, Dammit, Dampe98, Dan Tarrau, Dan100, Dan56, DanGalligan, DanKeshet, DanRF, DanShearer, DancingPhilosopher, DangerousBob, Dangoodman, Daniel, Daniel J. Leivick, Daniel Morales, Daniel Quinlan, Daniel4sw, DanielCD, Danieldis47, Danielfarzan, Danieljo2013, Danielsavoiu, Danigoni, Dankie, Danny 17, Danoasis, Dans tes rêves, Dante2308, DanteAgusta, Dantesqueman, Danthemankhan, Dar-Ape, Daredude, Dargen, Dark Shikari, Dark glaive, Dark12334312, DarkAudit, Darkfrog24, Darkness Shines, Darknshadow, Darkwind, Darrenemerath23, Darth Panda, Darthbob100, Dasani, Dave Lars, Dave Runger, Dave.e.bondeson, DaveSDCali, Daveharvey83, Daven200520, Daveswagon, David H Braun (1964), David Johnson, David Kernow, David Levy, David s graff, David.Monniaux, DavidFarmbrough, DavidLevinson, Davidcannon, Daviddamoore, Davideus86, Davidgrazi7, Davidmcarson, Davidstrauss, Davidyoung, Davila, DavisGL, DavyJones14, Dawn Eastwood, Dawnseeker2000, Db099221, Dbabbitt, Dbachmann, Dbenbenn, Dbiehl, Dbtfz, Dc freethinker, Dcammz, Dcandeto, Dcflyer, Dday14044, Ddlamb, Ddssaa, Ddye, De Moriarty, DeLear2012, Deacon of Pndapetzim, Deanb, Deanos, Deathmelon92, Deathperson, Deathregis, Debate789, Deborahhosen, Debouch, Decumanus, Deeceevoice, Deep impact, Deepstratagem, Deeptrivia, Deetdeet, Defective 13, DefendEurope, DeifieD, Delaywaves, Deli nk, Delirium, Delj523, Delldot, Delta x, DeltaSpectre, Deltabeignet, Demomoke, Demon-lad, Demonix, Den fjättrade ankan, Denelson83, Denihilonihil, Dennis44, DennisAro, Deor, Der4, DerHexer, DerMeister, Deranged bulbasaur, Derek R Bullamore, Derek Ross, Dereks1x, Descendall, Destructoman, Deturtlemon1, Deus Ex, DevastatorIIC, Devil1337, Deville, DevinCook, Dewan357, Dewy9, Dfarmer, Dffgd, Dfrg.msc, Dgilman, Dgw, Dharmabum420, Dhn610, Dhruvhemmady, Diaments 7.0, Diberri, Dickrichards, Diderot's dreams, Differentgravy, DigbyDalton, Digitalme, Dillan.Murray, Dillon29, Dilweg, Dina, DinajGao, Dinopup, Dipankan001, Dipper2, Dirtjacket, Discordanian, Discospinster, Dismas, Disneyland Resort4, Dispenser, Dissident, Distal24, Diverse82, DividedByNegativeZero, Dizzyizzy, DjDark1 2009, Djinn112, Dkdantastic, Dlohcierekim, Dlwldud96, Dmartin969, Dmeek7750, Dmeranda, Dmerrill, Dnheff, Dniete97, DobbyD, Doc Quintana, Doc glasgow, DocKino, Docu, Doczilla, Dogma5, Dogs2000, Dolemite17, Dom188816, Domaleixo, Dominic, Dominics Fire, Domminico, Domthedude001, Donarreiskoffer, Donsiku, Donut2, DoomsDay349, Dooney 30, Doopokko, Doprendek, Dorfner, Doringo, Dostluk, DoubleBlue, Doubleodd, Doug Funny, Doug Johnson, Dougdorval, Doughboy, Dougweller, Doverspa, Download, Dp462090, Dpaajones, Dpark, Dpm12, Dposse, Dpotop, Dpr, Dr kapp 89, Dr.alf, DrVerlucci, Draconas, Draconix, DragonCurse, DragonFlySpirit, Dragonfir731, DragonflySixtyseven, Dragonsscout, Dragoon1025, Draicone, Drakky, Dralwik, Dran0259, Dranzini, Drcornwallis1, Drdyer, Dreaded Walrus, DreamGuy, Dreamafter, Dreish, Dres, Drestros power, Driftwoodzebulin, DriveMySol, Drmies, Drtjumper, Druid1090, Drumguy8800, Drummingboi, Dsaddsa, Dsfsdf, Dthdc4, Dtobias, Dtwiggy, Dubayuw, DubbaG, Ducker, Ducknish, Dude018219293, Dudewheresmypizz4, Dudtz, Dufekin, Dukakis1988, Duke Hazzard, Dumaka, Dunro, Durin, Durova, Durruti, DustFerret, Dustimagic, Dustintml, Dusty777, Dutchdrummer99, Dvdcdr, Dwilso, Dwrcan, DylanW, Dynam1te3, Dynamicknowledge28, Dysprosia, Dzulco, Dzzl, Długosz, E Pluribus Anthony, E Pluribus Anthony redux, E30e, EH74DK, ENeville, ESkog, Eagle4000, Earl Andrew, Earpol, Earth, Eastlaw, Easyid, Ebakunin, Ebolito, Eclecticology, Economiesofscale, EconomistBR, Ecopirate, Ectoplasmer, Ed Poor, EdBever, Edcolins, Edgar181, Edgars2007, Edgeshappy, Edgeweb, Editor18, Editor19841, Editore99, Edmundbaird, Edobrichoo, Edsanville, Eduardo Sellan III, Edward, Edward Z. Yang, Edward nz, EdwinHJ, Edwy, Effer, Eggeggeggegg, Egil, Eileen's Dream, Eink21, Einzig27, Eivindgh, Ek8, Ekcrbe, Ekrub-ntyh, El C, El Generico, El Slameron, El kulu ngile, ElBenevolente, ElTyrant, Elbbom, Elduncarno, Electionworld, Electromud, Electrosoccertux, Elfguy, Eli81993, EliasAlucard, Elitistnerd, Ella Plantagenet, EllenCT, Elliskev, Ellomate, Elockid, ElockidAlternate, Eloquence, Elpato, Elpaw, Eluchil, Eluchil404, Emanuel Kingsley, Emcee2k, Emigrant85, Emijrp, Emilson, Emmette Hernandez Coleman, Emorph, Emperors Harbinger, Empion, Emsko, Emx, Encephalon, Encheart, Enciclopediaenlinea, EncycloPetey, Energizer07, Energyfreezer, Enkrates, Enter Movie, Entirelybs, Enviroboy, Eoghan888, Eoin123, Eorde, Epbr123, Epeefleche, Epicadam, Epicgenius, Epiovesan, Equilibrium007, Erachima, Erfemf, Erginer, Eric outdoors, Ericamick, Ericci8996, Ericwirtanen, Erikrespo, Erudy, Erwetr, Esimal, Eskimo Boots, Esrever, Essaregee, Esvbie\eongvrklakjmatthew, Eswiki, Etaoin, Etcher, Ethan c.00, Ettrig, Eu.stefan, Eugen Simion 14, EugeneZelenko, Euphoriaq, Eurocopter, Eurofire, Euroleague, Evans1982, Evb-wiki, Evelhal, Evensteven1, Evercat, Everyking, Evhs00, Evil Monkey, Evil saltine, Eviloldman, Evolauxia, Excel23, Excimer3.141597, Existensorg, Exlibris, Extra999, Extremalnik, Eyu100, Ezhiki, Ezratrumpet, F. H. Black, F20ljc, FAH1223, FCStaehle, FDR, FESTORM 5, FEastman, FF2010, FFLaguna, FOARP, FSF-Rapier, FT2, Fabartus, Fabhcún, Fabrictramp, Facts707, Fadeddreams5, Faelomx, Faezdel, Fagnostic, Failofbeaner, Faisal 1918, Faithlessthewonderboy, Falconclaw5000, Falcorian, FallingGravity, Fame, Fangz, FargoWells, FarnhamJ, Fastily, Fatbug1, Fatla00, Fatpeople12345, Favonian, Fctk, FeanorStar7, Fearthesun, Featuresaltlakecity, Fedallah, Feinoha, FelisLeo, Fenice, Fennec, FeralitYzERO, Ferio252, Ferkelparade, Fernirm, Ferrariman, Feydey, Feyer, Feysaln, Fezzaga, Fhavenvt, Fibo1123581321, Fidelpmn2n, Fieldday-sunday, Fiet Nam, Fifty7, Filby, Filemon, FinkleDude, Finoman40000, Firebug, Firehose, Firemancvn72, Fishal, Fishiehelper2, Fitzy2280, Fixman, Fkarim, Flackson, Flag-Waving American Patriot, Flamanchti, Flamarande, Flamealchemist, Flashflash;, Flatterworld, Flaunted, Flexure, FloNight, Floaterfluss, Flockmeal, Flogiz, Flooey, Flosssock1, Fluent aphasia, Flurry, FlushinQwnzNyc, Fnfd, Folksong, Following specific instructions whispered by a mysterious cat, Fonduelazone, Fonzy, Football123213, Footballfan190, ForemanDomai, Former user 6, Forsyth.joel, Fortdj33, Fourthords, Fowlance, Fplay, FraDany, Fragsta, FrancisTyers, Francs2000, Frangibility, Frank, Frank0051, Frank94.kane, FrankCostanza, Frans Fowler, Frazzydee, Freak in the bunnysuit, Freakofnurture, Frecklefoot, Fred Bauder, Fredbauder, Frederico1234, Fredericoxxx, Fredlyfish4, Fredrik, Fredy.00, Freepenguin, Frencheigh, FreplySpang, FreshBulletTime, Freyr, Freyr35, Frichmon, Friedo, Frietjes, Frjlove, Frogsprog, Frood, Frostyservant, Frullic, Frustrated american progressive, Fshoutofdawater, Func, Fundamental metric tensor, Funnyhat, Funsides, Furdeen, Future2008, Futurist110, Fuzheado, Fuzzygenius, Fuzzyslob, Fvincent, Fvw, G Rose, G. Campbell, G.W., G2bambino, G8summit, GCarty, GD, GHe, GOP904, GRBerry, GRuban, GSTQ, Gabbe, Gabrielsimon, Gadfium, Gaff, Gagi, Gailbawlz, Gaius Cornelius, GaiusTimiusAwesomus, Gaiusknight, Gakon5, Galati, Galoubet, Galwhaa, Gamaliel, Gamer007, Gamer9832, GamerMod, Gamersedge, Ganesha1, Ganeshk, Ganrither, Garfieldt, Garfld986, Garkeith, GarnetRChaney, Garret Beaumain, Garretgwozdz13, Gary King, Garynine, Garytmlee, Gasolene, Gator1, Gatsby4life, Gauss, Gaylencrufts, Gaz, Gazpacho, Gbd1995, Gclinkscales, Gdo01, GearedBull, Ged was here91, GeeZee, Geeoharee, Gegik, Gekaap, Gene Nygaard, GeneralDuke, Genghiskhanviet, Gengis, Genius Chimpanzee, Gentgeen, Geo Swan, Geo0910, Geo247, Geofat, Geoff Plourde, GeoffreyVS, Geoking66, GeometryGirl, George, George415, GeorgeFormby1, Georgeadgdgdgdwngo, Georgia guy, Geothermal, Gerald Farinas, Gerg1031, Germanlight, Gershwinrb, Getcrunk, Getoar TX, Gfnb, Gggh, Ggunit, Ghalas, Ghermit, Ghost1552, GhostPirate, Gianfranco, Giftlite, Gika, Gilderien, Gilgamesh, Gilhun, Gilliam, Gillydude, Gimferrer, Gimligimli1, Gimmetrow, GinaDana, Gingey933, Gintong Liwanag Ng Araw, GiovBag, Gjking, Gkmx, Gku, Glaucus, Glaxxo1, Glenn, Glenrqz, Globeism, Gman124, Gnaughton, Gnevin, Go229, Goalie1998, Gogo Dodo, Goheels36, GoingBatty, Gojoshg, Golbez, Gold Index, GoldRingChip, Goldblue21, Goldom, Gomarlins, Gomm, Good Olfactory, Good friend100, GoodDay, GoodSirJava, Goodandevil, Goodoldpolonius2, Goosehumper, Gop 24fan, Gopman1, Gordonfan, GorillazFanAdam, Goswamir14, Gr8opinionater, GraemeL, GraemeLeggett, GraemeMcRae, Graft, Graham87, GrahamTM, Gramaic, Grampion76, Grandin, Granpuff, Grant.Alpaugh, Graphic, Grapplequip, Graufbold, Green Giant, Green01, Greengreengreenred, Greenman, Greg96786, GregAsche, Gregcaletta, Gregorio, Greier, Greip, Grendelkhan, Gretzkyv99, Grey Shadow, Grey Wanderer, Greyhood, Grick, Grig541, Grimhelm, GrittyLobo441, Grjfiridkeokiski8j, Grm wnr, Grod028, Grop, Grosplant, Grosscha, Ground Zero, Grstain, Gruguuru, Grundle2600, Grunny, Grunt, Gsarwa, Gshave, Gsingh2300, Gsken, Gtuthtnfgurnfu, Guanaco, Guanxi, Gubbagubbaba, Guettarda, Gugilymugily, Guilherme Paula, Guitar 01, Gullit Torres, Gumbogumbo, Gundu1000, Gurch, Gurchzilla, Gurlpleezer101, Gurubrahma, Gustavb, Gutza, Guuzjo, Guy M, Gvhgu, Gwenham, Gwern, Gwernol, Gwickwire, Gwillhickers, Gwybedyn, Gypsum Miner, Gyre, H0ttamale2, H1478971, H5+R1A, HAHS 25, HIDECCHI001, HIZKIAH, HJ32, HTBrooks, HU Pilot, Habsfannova, Hackman911, Hadal, Hadrian89, Hadseys, Haemo, Hager jeff, Hahachuckles, Haham hanuka, HaireDunya, Hairy Dude, Haisthater, Hajor, Halberdier25, Halcionne, HalfShadow, Halibutt, Hall Monitor, Hallows AG, Hamedog, Hammer1980, Hans Dunkelberg, Happy-melon, HappyVR, Happyme22, Hardyplants, Harej, Hargrimm, Harkell, HarrietTubman, Harriganesh, Harro5, Harryboyles, Haruo, Harvestdancer, Hateless, Hateusa11, Haukurth, Haverberg, Havoc 8790, Hawaiian717, Hayden43, Haydenstone2, HazelGHC, Hazhk, Hcheney, Hchrishicks, Hcnorris, Hdsijhfoiuahfyusdaof, Hdt83, HeadCaptain, Headbanger756, Headlikeawhole, Heart Like a Wheel, HeartofaDog, Hearts of clear lake, Heat3000, Heaviestcat, Heby73, Hectorlf, Heegoop, HeikoEvermann, Heimdal, Heisencat, Hejsansvejsan, Hektossus, HeleneSylvie, Helios solaris, Helix84, Helixblue, Hellcat fighter, Hello14, Hello32020, Hello5959us, Helz0415, Henrik (old), Henry Delforn (old), Henry Flower, Henrygb, Hephaestos, Hequals2henry, Herdrick, HereToHelp, Hereforhomework2, Hernlund, Heroeswithmetaphors, Herp Derp, HerrGoebbles, Herunar, Herzie, Hexlor, Hfarmer, Hgspring, HiEverybodyHiDoctorNick, HiLo48, Highfields, Highvale, HillChris1234, Hillbillyholiday81, Himanis Das, Hipocrite, Hippiedude0, HippoOwl, HisForeShadow, Historian19, Historic Phoenix, Hiuby, Hk1992, HkCaGu, Hmains, Hmghosthost, Hobbentodd, Hobbesdude, Hobojoeandco, Hohum, Hoi jeon moo sool, Hojimachong, Hole62, Holledgeh, HollyAm, Holy Ganga, Homestarmy, Homeworld1031tx, HoodedMan, HooperBandP, Hoovernj, Hoping To Help, Hoponpop69, Hoppyh, HoratioVitero, Hordaland, Horses In The Sky, Hoshie, Hosterweis, Hot Stop, Hotelcalifornia1976, Hotshots2006, Hottentot, House1090, Househelps, Housesitter, Howard Alexander, Howcheng, Howdybob, Hseldon10, Hu, Huadpe, Huangdi, Hudude, Huduoroi, Huelenrite, Hugo Chavéz, Hugo999, Human diversity, Humanproject, Huphelmeyer, Hurricane111, Hurricanefloyd, Husond, Hut 8.5, HuuhManFui, Huzzahmaster018, Hydrogen Iodide, Hypathia, Hyperscope, Hystlen Caeth, Héous, I-2-d2, IANVS, IJA, ILguy06, ILuvAmerica, IMOW, IPodAddict181, IRelayer, ISpinksy, IZAK, Iakane49, IamBeastly20, IamYossarian, Iamthejake2000, Iamwisesun, Ian Dunster, Ian13, IanBailey, Iandiver, Ianhatch, Iankap99, Ibagli, Icairns, Iceberg3k, Icecradle, Iced Kola, Icedevil14, Id447, Idaltu, Idib, Ief, Ifrit 1780, Iggy402, Iggymwangi, Igilli, Ignacio Egea, IgnorantArmies, Igo4U, Ihatepotsmokinghippies, Ikanreed, Ikcyzrteip, Ike9898, Ikh, Ikyork, Ilikepie2221, Illegitimate Barrister, Illuminatiscott, Iluvmesodou, Ilya1166, Ilyanep, ImNotRichImStillLyin, Imaginary heroes, Imaninjapirate, Imbilly, Immunize, Impactplayer, ImperatorExercitus, Imperium Europeum, Improper Bostonian, Imzadi1979, In Defense of the Artist, InGearX, InShaneee, InTheRevolution2, Inanthropomorphism, Inchead, Indexheavy, Indopug, IndyLawSteve, Infidelis, InformationalAnarchist, Infrogmation, Inglok, Inhakito, Inimino, Inka 888, Innab, Insommia, Insuperable, Integralolrivative, Intelati, Intellibeing, Intelligentlove, Intelligentsium, Interfector, Internet quickie, Intimidated, Intmotinf, Introman, Intuitionz, InverseHypercube, Inwind, Ionutmortu, Iota, Iridescent, IrishHermit, Irishguy, Irishpyro, Irixman, Irrypride, Iruss14, Irviding11, Isawooty, Itai, Italia2006, ItalianGreyhound,

Italianboy10, Italiano111, Italo, Iudaeus, Iuio, Ivankinsman, Ivannoriel, Ivejustbegun, Ivirivi00, Ixfd64, Izehar, J Clear, J Di, J intela, J. Daily, J. Finkelstein, J.delanoy, J3ff, J4bin, JACOB911, JAHL14, JB50000, JBH23, JCRules, JDooley, JEN9841, JForget, JG2K6, JGHowes, JHMM13, JHP, JHarrelson, JHunterJ, JIP, JJARichardson, JLMadrigal, JLaTondre, JMBZ-12, JMS Old Al, JPArnett, JPD, JPMcGrath, JPaestpreornJeolhlna, JStewart, JTBX, JTM, JVz, JW1805, JWSchmidt, JZZDD, JaGa, Jaberwocky6669, Jabolba, Jacda1313, Jacek Kendysz, Jacj, Jack.Hartford, JackDayton, JackLumber, Jackbirdsong, Jackdude101, Jacketpook, Jacksoncw, Jacksonsimon29, Jacky man Toronto, Jacob Lundberg, Jacob.clarke19617, Jacobah, Jacobolus, Jacottier, Jadabocho, Jafd88, Jak2, Jaked92, Jakro64, Jalada, Jalfredl, Jambalaya, Jamcib, James A. Stewart, James.d.mccollum, JamesMLane, JamesMadison, JamesMaxwell, JamesReyes, JamesTeterenko, Jamesday, Jamesmarkhetterley, Jamespeterka, Jamie C, Jamiemason, Jammy-dodger, Jamyskis, JanCeuleers, Jane tyra, Janechra, Janilson1995, Janviermichelle, JanvonBismarck, Jaranda, Jarble, Jared mckittrick, Jarheadigor, Jarjarbinks10, Jarkeld, JarlaxleArtemis, Jasgbro, Jason Potter, JasonAQuest, Jasoncward, Jasondmath, Jasonfitz, Jasper Deng, Jatkins, Java13690, Java7837, Jawsman2, Jaxad0127, Jaxbobby, Jaxl, Jay, Jay942942, Jayafrancis, Jaydee928, Jayemd, Jayhawk08, Jayron32, Jazz214, Jazz457, Jbloun1, Jc900, Jcheckler, Jcm4, Jcmenal, Jcrocker, Jcw69, Jd027, Jdavidb, JdeJ, Jdforrester, Jdjoz0098, JeLuF, Jeancey, Jecar, Jecowa, Jedi master laybourn, Jeepday, Jef018, Jeff G., Jeff3000, JeffTL, Jeffgoin, Jeffrey S, Jeffz, Jenda H., Jengod, Jenks24, Jennavecia, Jennica, Jensboot, Jerahad, Jeremy221, JeremyA, Jeremygbyrne, Jeronimo, Jerry Jones, JerryFriedman, JerrySteal, Jerzy, Jess567, JesseRafe, JesseW, Jester2001, Jetlagorange, Jeyt5, Jez, Jezhotwells, Jezpuh, Jfarhou, Jfjkdc, Jfrancis, Jglees, Jgritz, Jguk, Jguk 2, Jh51681, Jhendin, Jhsounds, Jiang, Jibran1, Jihad!, Jim Douglas, Jim T, Jim1138, JimIrwin, JimWae, Jimbo D. Wales, Jimmy Kustes, Jimmy Slade, Jimmy butt, Jimmy-barnes, JimmySlade, Jimmyhead521, Jimothytrotter, JinJian, Jitse Niesen, Jivedevice, Jj137, Jj98, Jjacobucci, Jjb, Jjhcap99, Jkbena612, Jkfp2004, Jklin, Jkschroeder27, Jleon, Jlujan69, Jm51, Jmabel, Jman1974, Jmbrock, Jni, Joachimus, JoanneB, Jobe6, Jobroluver98, Jodafire, Joe Jarvis, JoeBlogsDord, JoeSperrazza, Joebieber, Joeblakesley, Joefromrandb, Joehaer, Joel27, Joelr31, Joelv40, Joesmoe8, Joesr55, Joeyramoney, Johan91, Johann Wolfgang, JohannVII, John, John Fader, John Price, John Quincy Adding Machine, John Smith M.D., Ph.D, John Torn, John-1107, JohnDBuell, JohnOwens, JohnRuskin, JohnWittle, Johndoe85839, Johnian144, Johnleemk, Johnmc318, Johnmike81, JohnnyMrNinja, Johnowenlangham, Johnstone, Johntex, Johntinker, Jojhutton, Joker123192, Jokeroler, Jon Doh, JonMoore, Jonarooni, Jonas Viper, JonasanRat7, Jonathan Drain, Jonathan-Morris711, JonathanFreed, Jonathunder, Jondel, JonesMama, Jongpil Yun, Jonnie d smith, Jonny Nixon, Jonsafari, Jonyyeh, Jooler, Joolz, Joowwww, Jophus00, Jorcoga, Jordanian forever, Jordz, JorgePeixoto, Jorgezoltan, Jose77, Joseph Buchanan, Joseph Solis in Australia, Josh Parris, Joshdboz, Joshua Scott, Joshua368, Joshurtree, Jossi, Josteinaj, Jowe189, Joy, Joylifen, Joyous!, João Neves, Jpers36, Jpgordon, Jprg1966, Jprulestheworld, Jrboi, Jrdioko, Jredmond, Jrkarp, Jrkso, Jrn, Jrobin08, Jrockley, Jros83, Jrt989, Jrtayloriv, Js00902, Jsoltman, Jtdirl, Jtkiefer, Jtokarsbr, Jts6293, JuJube, Juantay, Judgement, Judgesurreal777, Judygreenberg, Julain Barbarosa, JulieADriver, Jumbuck, Jumple, Jumpythehat, Junes, Jupiter Green, Juppiter, Jusdafax, Jusjih, Jusme1, Just a member, Justanotherguitarist, Jwissick, Jwking, K. Lastochka, K157, K50 Dude, KBtoys09, KConWiki, KF, KFan II, KGBarnett, KGasso, KL22, KS-U92, KTC, Kaasje, Kablammo, Kadelia, Kafziel, Kaio-ken x10, Kaisershatner, Kajal4, Kajmal, Kaldari, Kalsermar, Kampfers, Kangy, Karam.Anthony.K, Kardrak, Karimarie, Karl, Karl Dickman, Karukera, Kasyapa, Katalaveno, Katefan0, Kathmandu2007, Katydidit, Kaveh, Kazikame, Kazu-kun, Kazvorpal, Kbdank71, Kbolino, Kedar63, Kedi the tramp, Keebler71, Keegan, Keegster5, Keelm, Keilana, Keith D, Keith H., Keithd, Kellen, Kelly Martin, Kelstonian, Kelvinc, Kember, Kemet, Kemiv, Kemmabrant, Ken Seh, Kenaldinho10, Kennard2, Kenndo, Kennethduncan, KensterFox, Keon Milton, Kerrysfrench, Kertof, Kesla, Kesselring88, Kethinov, Kev2310, Kevin B12, Kevin Myers, Kevin Taylor, Kevin12xd, Kevinmon, Kevinp2, Kevjumba, Kevs, Kewp, Kg vasilev, Kgrad, Kgrr, Khajidha, Khalid, Khanada, Khatores, Khatru2, Khazar2, Kiddster93, Kier07, Kieranandersn, Kievaughn, Killdevil, Killer Swath, Kimon, Kinaro, King Dracula, King Gemini, King of Hearts, King wiston, KingBob1, KingTT, Kinggeedorah, Kingpin13, Kingturtle, Kinneyboy90, KinturkMan, Kirby of Doom, Kirchherr, KirchnerE07, Kire 1123, Kirill Lokshin, Kirkieboy, Kishorekumar 62, Kitch, Kitsunegami, Kjaer, Kkm010, Klaam, Klafubra, Klander Brigade, Klaser, Kliituu, Klonimus, Klr390, Kmacsimus, Kman543210, Kman665, Kmanblue, Kmaster, Knife2sharp, Knightshield, KnightxxArrow, Knorpel, KnowledgeOfSelf, Knowsitallnot, Knoxbury, Knutux, Koavf, Kochamanita, Kogsquinge, Konstantin, Korora, Kotjze, Kotniski, Koyaanis Qatsi, Kozuch, Kpjas, Kr1st1deejay97, Kraalg, Krakaet, KrakatoaKatie, Kralizec!, Krator, Krellis, Krich, Krismanx, Kristian Vangen, Krizpykremekid, Krupo, Kschwerdt514, Ktsquare, Kttreasure, Kude90, Kudzu1, Kueldude, Kukini, Kulikovsky, Kumioko (renamed), Kunaygi, Kungfuadam, Kurieeto, Kurt, Kurtle, Kuru, Kurykh, Kusma, Kuzco Mii, Kuzmaxster, Kwame1234, Kwamikagami, Kwijiborjt, Kwirky88, Kwizzard, Kxcd, Kyh87, Kyle Barbour, Kyle J Moore, Kyle sb, Kyle1278, Kyle4321, Kzzl, Kızıl Şaman, L33th4x0rguy, L3p3r, LGagnon, LIbertyInSpace, LOL, LaMenta3, Labby24, Labeledstickmen, Labz, Lachatdelarue, Lacrimosus, Lactical, Ladb2000, LadnaV, Laguna72, Lakshmix, Lala0000, Lalein91, Lamp coin, Lampoonsvacation, Lan56, Lanboney, Langston, Laomei, Lapsus Linguae, Lardy123, Lasindi, Last king of Frisia, LaszloWalrus, Latics, Latitude0116, Lauboo, Laubz83, Laucheehung, Laudaka, Lauren, Laurinavicius, Lawe, Lawrencekhoo, Lawyer2b, Laxbrain0110, Lazyandy123, Lcarscad, LeAviateur537, Lear 21, LeaveSleaves, Lec domicilarias, Lectonar, LedRush, Ledzeppelin321295, Lee Daniel Crocker, Leeannedy, LegitimateAndEvenCompelling, Lego872, Legoboy2000, Legokid, Legoktm, Lehi, Leif, Leithp, Lemonaderox, Lenoxus, Lenzar, Leodj1992, Leon7, Leonard G., Leroyinc, Leszek Jańczuk, Leuko, Levineps, Lewisgs, Lexor, Lfdder, Lfgnyc, Lfh, Lhademmor, Liber, Liberalcynic, Liberi Fatali37, Liechten11, LifeStar, Liftarn, Lightdarkness, Lightlowemon, Lightmouse, Lightningfro, Ligulem, Lijealso, Likeminas, Lilac Soul, Lilpinoy 82, Lima bean of the north, Limideen, LincolnSt, Lindberg, Lindenhurst Liberty, Ling.Nut3, Linuxbeak, Lipton1995, Lisa517, Lisazipper, LittleDan, LittleRoughRhinestone, Littlelance, Livajo, Living001, LizardWizard, Lizzie Harrison, Ljriccio9j, Llakais, Llec, Lloyd.smith, LlywelynII, Llywrch, Lmusielak, Lneal001, Locaracle, Lockesdonkey, Logan, Lokibolp, Lolnoraj, Londomollari42, LonelyMarble, LongBay, Longhorns4life89, Longwalkshortpier, Loodog, Looper5920, Looxix, Lord Emsworth, Lord Geznikor, Lord Moostyer, Lord Terminus, Lord Voldemort, Lord Wikipedian of the Wiki, Lord alexandre, LordHarrison, LordSimonofShropshire, LordSnooze, Lordflamel, Lordvaluemart, LoreleiLynn, Lorenzop, Lorenzosabbadini, Loser11111111122222, Lotje, Lou1986, LouI, Louis Waweru, Lousyd, Louturks, LovesMacs, Lowellian, Lowreeder, Lowride, LoyolaDude, Lrusso99, LtWinters, LuGiADude, LuK3, Luby, Lucas Duke, LudicrousTripe, Ludvikus, Luigi5256, Luiscolorado, Luisztdt, Luk, Luke334, Luke56, LukeM212, Lukemadness, Luna Santin, Lupin, Lupita°, Lupo, Lupus Bellator, Lussier, Luxor99, Lycurgus, LynnMeilleur, Lynntoniolondon, M a s, M2MM4M, M5891, MAG Fry, MBisanz, MC MasterChef, MCRTGP, MDCore, MECU, MER-C, MJCdetroit, MJDTed, MJEH, MJK, MKers06, MONGO, MPS, MWaller, MWatson15, MZMcBride, Mabbass, Mac Davis, MacGyverMagic, Macfan93, Macg4cubeboy, Macintosh User, Macjaeh, Mackay 86, Mackensen, Mackoy20, Macphisto12, Madbiggie00, Madcoverboy, Maddiekate, Madman445, Madskillz225, Maelnuneb, Maestro25, Magicman710, Magister Mathematicae, Magister Scienta, Magnet paper, Magog the Ogre, Maheshvenk, Maijinsan, Mailer diablo, Mailman9, Mais oui!, Majoreditor, Majorly, Makaristos, MakeChooChooGoNow, Malarious, Malcolm, Malhonen, Malo, Maltmomma, Mana Excalibur, Mandey7, Manicsurge, Manix, Manop, Mansmokingacigar, MantisEars, Manutdglory, MarSch, Marcfromns, Marcika, Marco Neves, Marco polo, MarcoTolo, Marcoescobar, Marcopolo7890, MarcusMaximus, Marcusliou5, Marcworca, Mareino, Marek69, Marielleh, Marine(uc), Mark, Mark Antony666, Mark Miller, Mark Renier, Mark Sheridan, Mark83, MarkDalit, MarkDeyarmond, MarkSweep, Markpravda, Marmote, Marnanel, Marrovi, MarsInSVG, MarsRover, Marshall Williams2, Marshmallow00, Marshman, Marskell, Martarius, Marth2k, Martin S Taylor, Martin451, MartinHarper, Martinense, Martinp23, Martinwilke1980, Martious, Martyman, Martynrss2, Marvolo Gaunt, Marymer, Marysunshine, Mashboy, Mason.Jones, Massimo Catarinella, Master Jay, Master and Commander, Master of Puppets, Master of the Oríchalcos, Mastercomputerpro9999, Masterjamie, Masterpiece2000, Matchgame12345, Mateo SA, Materialscientist, Materic123, MathewDill, Mathgod333, Mathpianist93, Matsaball, Matt 8, Matt Crypto, Matt Gies, Matt Hales RDFC@hotmail.com, Matt Heard, MattPDG, Mattabesta, Matthewprc, Mattkidd12, Mattt296, Mattworld, Matty 1988, Matty.007, Mattyh190, Mattyhar, MaulYoda, Mav, Max rspct, Max(is not)well, Max20020012002, Maxatl, Maximus Rex, Maximuspower, Maxkoyama, Maxkramer, Maxmc, Maxpower37, Maxsht9, MaxxFordham, Mayumashu, Mb93j, Mblumber, McBenjamin, McDogm, Mcaleppo86, Mcattell, Mceder, Mchcopl, Mcimo88, Mckyj57, Mcmoor, Mcsweet, MeThOd, MeasureIT, Meco, Medevila, Medo4, Medo9, Medvedev40, Meegar, Meelar, Meeptastic, Meeve, MegX, Megaandroid30, Megahmad, Megan 189, Megapixie, Megawattbulbman, Mehicdino, Melonkelon, Melsaran, Member, Menchi, Mendel, Merbabu, Merlot70, Mermaid from the Baltic Sea, Merovingian, Mesgul82, Messiisking, Metafisyx, Metehan1992, Mets501, Metz2000, Meursault2004, Mevlüt Kılıç, Mexihcatlacatl, Mfwnoface, Mghabmw, MghtyRoach, Mgiganteus1, Mh10190, Mhesselb, Mhking, MiKellie, Michael Essmeyer, Michael Glass, Michael Hardy, Michael Patrick, Michael of the World, Michael-Billa, Michael.Urban, Michael2-, MichaelBillington, Michaelenandry, Michaelpkk, Michaelwuzthere, Mickey gfss2007, Mickeykozzi1, Mickeymickey11, MicronationKing, Midlothian09, Midnight Green, Mightymights, Mihas, Mika293, Mikael Häggström, Mike Christie, Mike Halterman, Mike Pehipol, Mike Rosoft, Mike nofx, MikeBravo, MikeLynch, Mikefzhu, Mikehawk, Miken32, Mikeroodeus, Mikestern2, Mikeybadass05, Mikhail Ryazanov, Mikhailov Kusserow, Mikinator, Mikpass1, MilFlyboy, Mileage, Milk of magnesia, Milk the cows, Milkmandan, Millahnna, Million Moments, Millsdavid, MindscapesGraphicDesign, Minesweeper, Minesweeper.007, MiniZiper, Minjamonkeyw00t, Minna Sora no Shita, MinnesotanConfederacy, Miranda, Mirrorkirby9062, Mirv, MisfitToys, Miskin, Miss tylie, Miss-jessie-gal, Missionary, Missmarple, Missouri Rebel, Mistakefinder, Mister Alcohol, Misterrick, Mistertheman, Misza13, Mitaphane, Mitchumch, Mitsukai, Mitth'raw'nuruodo, Mixcoatl, Miyokan, Mizuphd, Mjf3719, Mkweise, Mlazik, Mm40, Mmarch31, Mmavipc, Mmx1, Mmxx, Mnlira013, Mnmazur, Mo0, Moby-Dick3000, Moderate2008, Modest Genius, Modulatum, Moe Epsilon, Mogden, Mogism, Momentary, Momo san, Momus, Mon amour est avec la France, Moncrief, Mongoletsi, Monkeyblue, Monkeyman, Monkeywrenchjim, Mono, Monobi, Monocletophat123, Monsieur Fou, Monsterhead150, Montrealais, Montyofarabia, Moondog88, Moonraker0022, Mooseberry, Mordien, Moreau36, Morgankevinj huggle, Morhange, Morphh, Morriswa, Morton devonshire, Morwen, Mosca, Mosemamenti, MosheA, Mostgiganticlyingmouthofalltime, Mosthauntedjp, Mostly Rainy, Motthoangwehuong, Mouloud47, Moustan, Moverton, Moxy, Mozartmuzic, Mprudhom, Mpscan, Mr Adequate, Mr Stephen, Mr Taz, Mr Twix, Mr. D Vega, Mr. Lama, Mr. bwn, Mr.Do!, MrD9, MrDrew508, MrGRA, MrJanitor1, MrMick, MrRadioGuy, Mrbelvedereposter, Mrbrown, Mrcolj, Mrdthree, Mrjpb104, Mrknowitman, Mrld, Mrmaroon25, Mrtobacco, Mrwojo, Mrzaius, Mschiffler, Msludwick, Msundqvist, MtnBrook, Mufka, Mulad, Mumble45, Munulu999, Muraad kahn, Murali nmv, Murchy, Murdocc, MurphysLawn, Musdan77, Mushroom, Musicalantonio, Musiphil, MuskokaON, Musse-kloge, Mustanglover, Muya, Mwanner, Mx3, Mxcatania, Mxn, Myanw, Mydreamistofly, Mygeneration01, Myleslong, Mylittlezach, Myriad, Mystaker1, Mystify85JEC, Mário, Mäximus Rex, N-edits, N-true, N.M.Sheedy, N00bschoolbus, N244888, N5iln, N734LQ, NBS525, NByz, NDSteve10, NGerda, NHRHS2010, NJOrlando, NJRobbie, NJguy281, NKT45, NSLE, NSR, NW036, NWOG, NYCJosh, NYKevin, Nach0king, Nachomania, Nadavspi, Naddy, Nader85021, Nafield, Nakon, Nakul Dhoot, Nappyrootslistener, Narco, Nareek, Naryathegreat, Nascar1996, Nat, Nat Krause, Nat682, Nate1012, NateXMS, Natgoo, Nathan8225, Nathanalex, Nathanbolt, Nathanshinsen, NativeForeigner, Nattuvijay76, Natty4bumpo, NatusRoma, Naus, Nautilator, Naval Scene, NawlinWiki, Nbatra, Nblc, Nbrzycki91, Nd24, Nduhn101, Neal Finne, Nebulousity, Nehrams2020, Neilbeach, Nelandquinten, Nelfer, Nelro, Nelsondavidd, Neo233, NeoLamagra, Neoballmon, NeonNight, Neondiscogod, Neorunner, Neovu79, Nerdalert101, NerdyScienceDude, Nerothanes, Netbenefit, Netesq, Netkinetic, Netoholic, Netscott, Neuroscientist, Neutrality, Nevada, Nevilley, New England, NewEnglandYankee, Newstormer, Nezbie, Nhakimi, Niceguyedc, Niceley, Nichalp, Nichofas, Nicholas F, Nicholas Tan, Nick, Nick.mon, Nick123, Nick4gwen, NickMartin, Nickptar, Nickyaberdeen, Nico, Nicolas Barbier, Nicolewasherexx, Nicolharper, Nidam, Niemti, Nige, Nigeria01, Nighend, Nigholith, NightDragon, Nightscream, Nihonjoe, Nika, Nikkimaria, Nikodemos, Nikofeelan, Nikola Smolenski, Nikoniko, Nikostar, Nikwad244, Nil Einne, Nil105994, Nimbley6, Nine999999999s, Nipplemaster5000, Nippoo, Nirvana888, Nishant12, Nishishei, Nissi Kim, Niteowlneils, Nitharsank, Nitya Dharma, Nivix, Njoedits, Nkrosse, Nmpenguin, No Guru, No toco at aol dot com, Noleander, Nomad, Nomist, Nonenmac, Noozgroop, NorCal764, NoraBG, Northamerica1000, Northmeister, Northwales4u, Northwest, NortyNort, Norwikian, Nosimplehiway, Nosskyline, Not a dog, NotAbel, Nothlit, Nottheking, NovaNardis, Novidmarana, Novis-M, Nowhither, Nrafter530, Nrana99, Nroscha, Nscheffey, Nskrill, NuKkEm, NubKnacker, Nuberger13, NuclearVacuum, NuclearWarfare, Nufy8, Nuge, Nukebros, Numberoneconti, Numberonesurvivor, Numerouno13, Nunh-huh, Nvw, Nxxos, Nydas, Nzzl, OBrienMadHouse, ONEder Boy, OOODDD, Obaidz96, Obhave, Obi-WanKenobi-2005, Obli, Oblivious, Occono, Octane, Oda Mari, Offenbach, Offrecord09, Ohconfucius, OhitstragikX, Ohmygodeven, Ohnoitsjamie, Okiefromokla, OlEnglish, Olaf Eriksson, Old Guard, Old Moonraker, Old Right, OldRight, OldRightist, Oldag07, OldakQuill, Oldmankdude, Ole10589, OleMaster, Oleg Alexandrov, Olegwiki, Olivier, Ollieegg, Olorin28, Omarlegend, Omarmorek, Omega9380, Omicronpersei8, Ommnomnomgulp, Omnibus, Once in a Blue Moon, One, One man orgy, OneDoubleO, OneEuropeanHeart, OnePt618, Onetwo1, Onixz100, Onopearls, Onorem, Oo64eva, Oobopshark, Opaldragon1, Opelio, Openthedoor754, Operation Spooner, Optim, OptimumPx, Optional Toaster, OrbitOne, Ordealbyfire, OreL.D, OregonD00d, Oreo Priest, Oreotoast, Orthorhombic, Oscarthecat, Oshwah, Oskart, Ost316, Otashiro, Otebig, Otolemur crassicaudatus,

OuroborosCobra, Outback the koala, Outoforderuk, Ovcp1, Owen, OwenBlacker, OwenX, Oxboomer, Oxford Comma, Ozi, P.W.Lutherson, P0per, PAWiki, PBS, PCock, PDH, PFHLai, PL290, PRRfan, PRSteve27, PS2pcGAMER, PZFUN, Pabejar, Paco8191, Pah246, Pahistorian, PainMan, Paine Ellsworth, Paix Pan, Pajz, Pak21, Pakaran, Paleo dragon, Paleoblues, Pamri, Panfakes, Panserbjorn, Pantherskin, Papasmurf0810, Papercrab, Para, Parachnidox, Paradise7, Parallel or Together?, Paraside, Parliamentcity, Parmadil, Parthasarathy B, ParvatiBai, Pass a Method, Pastorsson93, Patapsco913, Patioco, Patka, Patman2648, Patrick, Patrick19, Patrickneil, PatrikR, Patsson77, Pau07, Paul A, Paul August, Paul Tyrrell, Paul.h, Paul2.0, PaulHanson, Paulinho28, Pauljeffersonks, Paulmcguane, Paulmec, Paulsprecker, Pax:Vobiscum, Pb30, Pbedford, Pcman22594, Pcpcpc, Pcstico, Peak, Pearle, Pecher, Pedant, Pedantic Turkey, Peeweeteehee, Pegasus1138, Pekin Republican, Peloneous, Pendragon5, Pengkeu, Penny1020, Penser, Penshurstpride, Penubag, Pepesia, Pepper, Pepsi132, Pepsidrinka, Per Honor et Gloria, Peregrine Fisher, Perfect Editor, Perfect77, PerfectStorm, Perfectionaintperfect, Perkinator 3000, Permethius, Perseus, Son of Zeus, Persian Poet Gal, Peruvianllama, Pestcamel44, PeteJayhawk, Peter, Peter L, Peter.C, PeterK, Peterale4444, Peterlin, Pethr, Petrb, Petrux, Pewwer42, Peyman Ghasemi, Pfortuny, Pgan002, Pgecaj, Pgk, Ph.D.Nikki, PhJ, PhaedoHD, Phamdangkhoa2410, Phantomsteve, Pharaoh of the Wizards, Pharos, Phe, Phil Boswell, Phil5329, Philanthropy, Philip Stevens, Philip Trueman, Philip200291, Phillies1fan777, Philliesfan035, Phillip Fung, Phillip J, Philpill691, Phinnaeus, Phoebe, Phoenix2, Phoenix7777, Photo me2004, Phrankle, Phrasia, Phxcesar, PhysicsPat, Pi399, PiMaster3, Picapica, Pidgeontoes, Pierco, PierreAbbat, Pigman, Pilko182, Pillow2011, Pilotguy, PinchasC, Pine, Pinotgris, PinoyWikipedia, Pioelad, Piotrus, Piss420, Pizzadeliveryboy, Pk10534, Pkpat2011, Plainsong, Plapre09, Plastic, Plasticbadge, Plusgood, Pm06420, Pmackarate, Pman31014, Pmanderson, Pnatt, PoccilScript, Poetaris, Poilil, PokeHomsar, PokeYourHeadOff, Pol098, Polaron, Polarscribe, Polisci, Politepunk, Political hack, Politis, Polylepsis, PonileExpress, Poojean, Poopymatt, Poor Yorick, Poppa Bear, Porkman385, Porsche997SBS, Portillo, Postco, Postdlf, PotentialDanger, Potnpan123, Pperos, Prasenberg, Pratheepps, Premeditated Chaos, Preschooler.at.heart, Preslethe, PrestonH, Pretty Green, Preuninger, Primalchaos, Primarypenguin, PrinceForte, Prision, Pristino, Prjrv787, Prodego, Producercunningham, Prospect1, Proteus, ProudIrishAspie, Pryde 01, Pschemp, Pseudo-Richard, PseudoSudo, Psy guy, Pubber, Publicus, Puertorico1, Pulmonaryarchery75, Pupitar13, Purodha, Pushbreak, Pwilk, Pwjb, Pwnage8, PxT, Pyroclastic, Pyroechidna1, Pyukman, Pz821, Pé de Chinelo, Q Ramona, Qasaqsuyu, Qmwne235, Qqqqqq, Qrc2006, Quac, Quadel, Quadell, Quadrochosis, Quaker24, Quako5, Quanticles, Quasipalm, Quazywabbit, Quberoot, Query, Questchest, Quijebo, Quill, Quinn C56, Quinnculver, Quite vivid blur, Quixoto, Quizimodo, Quizimodo (usurped), Quota, Qviri, Qwe, Qwerasdfzxcv1234, Qwfp, Qxz, R'son-W, R-41, R-Joe, R0xorz, R9tgokunks, RA0808, RG2, RIPSAW1986, RJII, RJN, RMHED, RPIRED, RPlunk, RPlunk2853, RS085, RTG, RVD, RVJ, RWhite, RZ heretic, RaCha'ar, Rab777hp, Rachaelho50, Radarsnobles, RadicalBender, Radicalsubversiv2, RadioBroadcast, RadioKirk, Raegis, RafaelG, Rageinacage, Ragib, Rahlgd, Rahuldharmani, Rail88, Railer 893, RainR, Rainpat, Raistuumum, Raja Hussain, RajaNeela1993, Ral315, Ralph3000, Rama, Rama's Arrow, Ramanglass544, Ramazan!, Ran, RanEagle, Ranainam5343, RanchoRosco, RandalSchwartz, Rande M Sefowt, Random contributor, Randomblue, Randombob12345, Randomengine, Randy Johnston, Randy Kryn, Rangermike, Rangerwave, Rangoon11, Ranveig, Raoul-Duke, Raprchju, Rarelibra, Ratemonth, RattleMan, Raul654, Ravedave, Ravelnine, Ravenhull, RaviC, Ravikiran r, Ravlo2004, Raybob95, Rayc, Raymond arritt, Rayoflight278, Rb2911, Rbankson, Rcharding, Rctbone, Rd232, Rdrgz93, Rdsmith4, RebaFan1996, Rebellion101, Reboot, Reck345, Red Slash, Red rawker, Red65, RedChord, RedWordSmith, Redd Baron1, Redd Foxx 1991, Reddi, Redeem, Redgecko5, Redkloud, Rednaxela, Redwolf24, Reeces904, Reedy, Reelcheeper, Refdoc, Regancy42, Regionalsimp, Reid A., Reid1867, Reillyg, Reimelt, Reinthal, Reisio, Rejs12345, RekishiEJ, Reliancepowercoin, Rembo74, Remiel, Renegadeviking, Renorei, Rentastrawberry, Renzo92x, Republicanbetter, Republikaner, Reticulum, Retired username, Retireduser1111, Retrospector87, Rev.Ision, Revmachine21, Revolución, Rex Germanus, RexImperium, RexNL, Rextrim, Reywas92, Rfc1394, Rgriga1, Rhaegar, Rhatsa26X, Rhobite, Rholton, Ric36, Rich Farmbrough, Richard Allen, Richard Blatant, Richard David Ramsey, Richard0612, RichardDavila, RichardF, Richie123098, Rick Block, RickK, Rickipedia127, Rickney, Rickvaughn, Ricky@36, RickyTheBest, Rickyrab, Riconoen, Rifecu, RightCowLeftCoast, Rillian, Rimbaud, Riphamilton, RiseAbove, Risker, Ritchy, Ritimakovej, Rittmeister, Rivertorch, Rjairam, Rjd0060, Rjd2 Deadringer, Rje, Rjensen, Rjm656s, Rjwilmsi, Rklawton, Rkt2312, Rlevse, Rlm0710, Rlundi86, Rmerlos, Rmhermen, Rmhs15, Rmky87, Rob 06-07, Rob Shepard, RobTranter, Robbiemt, Robbo chappers, Robbro7, Robchurch, Robdurbar, Robert K S, Robert.Mcleod, Robert25, RobertG, RobertHuaXia, Robertkeller, Robertson-Glasgow, Robertvan1, Robfergusonjr, Robin63, Robster1983, Rocafella, Rocastelo, Rocawear1994, Rochabm, Rockentui, Rockfang, Rockstone35, Rockyobody, RoddyYoung, Rodrigogomespaixao, Rofl, Roger Arguile, Roger Gianni, RogerK, Rogered, Rogertudor, Rogueagent123, Rohawn, Roke, Roleplayer, Rollercoasterlord, Rollo44, Rolo75, Romanm, Rompe, RonH, Ronaldmolina20, Ronster21, Ronthewizard1994, Roosevelt Churchill, Rory096, Rory2259, Ros24s, Rose Garden, Rossenglish, Rosshterry, Rossnixon, Rotten, RoundDaWorld, Route11, RoverTheBendInSussex, Rox Da Box, Roxi2, Roxorzursoxorz, Roy Brumback, Roy da Vinci, RoyBoy, Royboycrashfan, Rreagan007, Rrhain, Rrie, Rrius, Rror, Rrouse, Rsm99833, Rstrug, Rto and buddy, Rtyq2, Rubber ducky222, Ruchi tspl, Rudjek, Rudolph.cm, Ruhrjung, Rumkles, Run26.2, Ruy Lopez, Rwalker, Rwenonah, Rwolfe85, RxS, RyGuy17, Ryan032, RyanGerbil10, RyanParis, Ryanbowden, Ryanjunk, Ryanluck, Ryanmcdaniel, Ryland510, Ryulong, Ryz05, RĭckK, S.U.H.E.L., S2006causa, S4diStiC, S71elements, S91by, S@bre, SCEhardt, SDJ, SDY, SEWilco, SFrank85, SG, SGBailey, SJMNY, SMC, SNIyer12, SPQRobin, SPUI, SS Division Wiking, SSZ, STARWARSGEEK, SWBJACMK, SWalkerTTU, SYSS Mouse, Sa.vakilian, Saashaw, Saathoff37, Saberswordsmen1, Sabrown100, Sabs1285, Sadads, Sadasd, Sade, Sadow300, Saforrest, Sagaciousuk, Sahrin, Saigon76nyc, Saksham, Sakutdjica, Salsb, Sam Coutu-Oughton, Sam Korn, Sam1993, Sam3199, SamEV, SamForestell, Samantha555, SameOldSameOld, Samir, Sammo, Sampo Torgo, Samroundface, Samuel Blanning, Samuel Sol, Samuelsen, Samwb123, Sanantonioguy, Sandersonjoe229, Sangjinhwa, Sango123, Sangre viento, Sannse, Sansvoix, Santiagobernstein, Sarah, Sardanaphalus, SarekOfVulcan, Saric, Saros136, Sasquatch, Sauerjos, Sauve.d, Savefrance, Savidan, Sayden, Saypotatoes8, ScAvenger, Scammerboy227, Scanlan, Scarab12, Scarian, Sceptre, Schadenfreude52, SchfiftyThree, Schmloof, Schnitzelman, Scholarly Breeze, Schoolsurfer, Schrödinger's cat, SchuminWeb, SchutteGod, Schutz, Schzmo, ScientificDude, Scientiom, Scientizzle, Scientus, Scientz, Scifiintel, Scifiradioguy, Scipius, Scirocco6, Sciurinæ, Scohoust, Scoobycentric, Scorchy, Scorpi0n, Scotchy500, Scott Burley, Scott Illini, ScottMainwaring, Scottmsg, ScriptBlue, Scrumy21, Scuppers1, Sdgjake, Sdhonda, Sdnegel, Seablade, Sean, Sean Whitton, Sean gorter, SeanO, Seaphoto, Search4Lancer, Searchme, Seb az86556, Sebbe, Secret (renamed), Secret Saturdays, Secretlondon, Seeraamaazu, Sekolov, SemDem, Semiquaver, Sennsa, Senorquigles, Separa, Sepeople, Seqqis, Seqsea, Seresin, Seric2, Serlin, Serminigo, Serpentinite, SerpentsTail, Sesel, Sesu Prime, Seth Ilys, Settl746, Sevillano1, Sexxie grrl, Sexy sally, Señorplayero, Sfacets, Sfahey, Sfan00 IMG, Sfmontyo, Sgguitarist, Sgodth, Shade11sayshello, ShadowGuardian, Shadowboy667, Shadowjams, Shadowy Crafter, Shaggy2Dope, ShakataGaNai, Shaken Earth, Shalom93, Shame On You, Shanedidona, Shanel, Shaneomac8605, Shanes, Shanesterman, Shannon7mcnamara, Shared Thought, Sharkticon Guy, ShaunES, Shawn m r, Shearonink, Shech736, SheikYerBooty, Shelleycakes, Sherzo, Shii, Shim jb, Shimonnyman, Shindleria235, ShipperS7, Shizhao, Shmew, Shoaler, Shoeofdeath, Shohei Okada, Sholtar, Shootappliances, Shoreranger, Shorne, ShortShadow, Shot007, Shpilo, Shredder2012, Shreshth91, Shroudan, Shunt110, Shushruth, Sicamous, Sicilianshotgun, Sietse Snel, Sig0, SigPig, Sigma 7, Sigmaparadox, Signsolid, Silence, SilentOpen, Silver Sonic Shadow, Silverchemist, Simhedges, Simishag, SimonP, Simulcra, Sinmiedoanada, Sir Nicholas de Mimsy-Porpington, Sir smorgasbord, Sir48, SirGrant, Sirkevinalot, Siroxo, SivaKumar, Sj, Sjakkalle, Sjö, Sk.Sajjad, Skaakt, Skapur, Skarebo, Skate4me, Skateboarderf2, Skeeter08865, Skentelbery, SkepticBanner, SkerHawx, Skier Dude, Skizzik, Skomorokh, Skooing, Skullfiend, Sky Divine, Skydeepblue, Skyking360, Skylinegtrgsl, Skyoon, Skyring, Sladuuch, Slakr, Slashcoconut, Slatersteven, Slconnor uk, Slcut, Sleptrip, Sligocki, Slivertransplant, Sloman, Slowking Man, Slrubenstein, Slugmaster, Slurpy121, Sm8900, Smaug123, Smcmanus, Smilemean, Smiling Kevin, Smiloid, Smith03, Smith2006, Smoke., SmokeyVillain, Smokeythecat, Smr16, Smug Irony, Snailwalker, Snoozer347, SnowFire, Snowmanradio, Snowolf, Snoyes, SoMuchTime, Soakologist, Soccerisawesome, Sodzei, Sole Flounder, Solipsist, Solitude, Somatosis, Some jerk on the Internet, Somedifferentstuff, Someone else, Somercet, Somethingoranother, Sonicfire3000, Sonjaaa, Sonyray, Soooclose, Soosed, Sopher99, Sophie2895, Sopranosmob781, Soulpatch, Soulstice, Soumyasch, Soundslikealotofhooplah, South Bay, SouthernComfort, Southsydney, Soxwon, Sp33dyphil, SpNeo, Space25689, SpaceFlight89, Spaceboy492, Spangineer, Spark98, Sparrowman980, Spartan078, Spartanloser, Spebudmak, Speedboy Salesman, Speedy06car, Spellmaster, Spencer195, Spettro9, SpiderMMB, Spiesr, Spiff, Spike105, Spikethehawk, Spinoff, Spiralhighway, Spitfire19, Spitzl, Splash, Splat, Spliffy, Sportsjam3, Spudtater, Spundun, SpuriousQ, Squash Racket, Squideshi, Squoups, Sri 1988, Srikantkedia01, Srishti1984, Srithikdatta, Srs, Srujanjoshi, Srushe, Ss2343, Ssilvers, Ssolbergj, Ssssss340, Sswonk, St00pidmonkey72, Stabby Joe, Stallions2010, Stan Shebs, StanZegel, Staplegunther, StarScout0304, Stars999, Starwrath, Stasven, Statalyzer, StateCollege101, Static11, Statsagogo, Staxringold, Stdevin2012, Steel, Stemonitis, Stenen Bijl, Stephen Gilbert, Stephenb, Stepshep, Stevage, Steve the Beach, Steve03Mills, Stevec240, Steven J. Anderson, Steven407653, Stevertigo, Stevietheman, Stevo1000, Stickee, Storasyster, Stormie, Stormscape, Strangepalefighter, Stratman07, Streryan, Strideranne, StringRay, Stroganoff, StrongProudGunny, Strongbad1982, Strovolos01, Stubblyhead, Student7, Stuman12, Stumink, Stupidstupidstupid, Stwalkerster, Subseven26, Substancep, Subsurd, Success123$, Sudeepneelakantan, SudoGhost, Suffusion of Yellow, Sufi dog, Suisui, Sumahoy, Sumanch, Sumone10154, Sunil060902, Sunquanliangxiuhao, Sunray, Sunshineofyourlove12, Supaman89, SuperDude115, SuperSmashBros.Brawl777, Supercoop, Supergeo, Supergodzilla2090, Superm401, Supersexyspacemonkey, Supertask, Superzoulou, SupremeRuler, Surcouf, Surgemg, Surlyduff50, Suruena, Suryoye, Susurrus, Suto, Sverdrup, Swainstonation, Swamilive, Swarm, Swatjester, Swedish fusilier, SwedishConqueror, SwedishZeta, Sweetback, Swellman, Swid, Swimm212, Swissbrazilian, SwnyMac, Swollib, Swordhunter0690, Syndicate, Synthe, Syrthiss, Sysrpl, Systemshocked, Syvanen, Szajci, Szakharin, Szyslak, T-rex, T1980, T800m101, TBrandley, TEG24601, TEMcGee, TFNorman, TFOWR, THEemu, THEunique, TJ Spyke, TJDay, TJive, TMC, TOMMASO, TOO, TORR, TR5000, TSO1D, TShilo12, TTBMA, TUF-KAT, TV1981HS, TXLonghorn, Ta bu shi da yu, TaRiQ 8d8, Tabletop, Taborgate, Tachyon01, Taco325i, TacticalMouse, Tad.k.22, Tafinucane, Tahir mq, Tai112, Taifarious1, Taikodrumfun, TakuyaMurata, Tamara Ustinova, Tangotango, Tanishh, Tannin, Tapir Terrific, Tapir2001, Tapirs65, Tardis, Targeman, Target Jackson, Tarheelz123, Tariqabjotu, Tarret, Tasc, TastyCakes, Tavatar, Tawker, Tayyabbaig, Tazchook, Tbhotch, Tbone55, Tbonnie, Tbsea, Tburgtbone06, Tdadamemd, TeaDrinker, Team6and7, Teammm, Techdawg667, Techguy118, Technical 13, Technocratic, Teddey, Tedickey, TehPlaneFreak, Tehmadhatter, Tekto9, TeleComNasSprVen, Teletoonmaniac, Telex, Tellsgeschoss, Telsa, Telstar2626436, TempDog123, Tempodivalse, TenOfAllTrades, Tenunda, Terence, Terfili, Terlynn4, Terra Novus, Terrasidius, Teryx, Testes rodriguz, TexasAndroid, Texaswebscout, Texinian, Texink, Texture, Tezero, Tf157, Tfine80, Tgannon, Tgeairn, Tha Nad 2007, Thadkobylarz, Thane, Thargor Orlando, Tharnton345, That-Vela-Fella, Thatdog, Thatotherperson, Thatpacokid, The Catholic Knight, The Cunctator, The Disco King, The Epopt, The Evil IP address, The Evil Spartan, The Four Deuces, The Great Liberator, The Halo, The Hitch22, The Land, The Lizard Wizard, The Man in Question, The Minister of War, The Number Man, The Ogre, The Placebo Effect, The Punk, The Revert States of America, The Tetrast, The Tom, The Transhumanist, The Universe Is Cool, The Utahraptor, The Wingman Joker, The ed17, The last Wikibender, The lost library, The machine512, The monkeyhate, The slave to reality, The ultimator, The wub, The-Holy-Hand-Grenade-Of-Antioch, The1edit, TheArmadillo, TheDoctor10, TheFarix, TheFuzzyFive, TheGWO, TheGrappler, TheHoosierState89, TheKMan, TheLongTone, TheMadBaron, TheNewPhobia, TheObtuseAngleOfDoom, ThePenciler, TheRanger, TheRomanLegion, TheThomas, TheTimesAreAChanging, TheVirginiaHistorian, TheWama, Theboondocksaint, Thebrother, Thechinese8, Thecrazydonvingigante, Theda, Thedec, Thedragonking, Thegargantua, Thegreatgrabber, Thegryseone, Thejason, Thelacker, Thelema418, Themane2, Themanwithoutapast, Themanwiththefish, Thematt523, Thenderson1, Thepebble3, Theredhouse7, Therequiembellishere, Theresa knott, Theriusthechronicler, Thermionic1, Thernlund, Therock40756, Thesevenseas, Thesouthernhistorian45, Thethirdperson, Theurgist, Thewolfstar, TheyMadeMeDoIt, Thezirk, ThievingBeagles, Thingg, ThinkBlue, ThinkEnemies, Thisbites, Thiseye, Thomas Paine1776, Thomas S. Major, ThomasK, Thomasjamesfoster96, Thomasmallen, Thomasmatoushek, Threepwood89, Thricecube, Thryduulf, Thunder8, Thunderbird2, Thunderboltz, Thurifer, Tide rolls, TigerShark, Tigey, Tightguy400, TigranTheGreat, Tim Starling, Tim Thomason, TimBentley, TimShell, TimVickers, TimothyHorrigan, Timrem, Timrollpickering, Timwi, Tintero, TintininLisbon, Titch Tucker, Titoxd, Tiwonk, Tiësto, Tjl1128, Tjmayerinsf, Tjoe, Tjpark1111, TkerTimeSeeker, Tktktk, Tktru, Tkynerd, Tlogmer, Tma 88, Tmazhindu, Tobby72, Tobycat, Tobyk777, Tocino, Tom, Tom Reedy, Tom harrison, Tom-, Tom8866, TomStar81, Tomass288, Tombyt55, Tomdatom, Tomgibbens, Tominator93, Tomlillis, Tommybtennis, Tommybtennis2, Tommyjb, Tomwood0, Tomwsulcer, Tony Sidaway, Tony age, Tony1, TonySt, Tonye104292, Tonywiki, Toonlad9, TopAce, TopazSun, Topbanana, Toropop, Tos, Tosfan4ever, Totla, Toussaint, Tparkes, Tpbradbury, Tpetross, Traal, Tragicfame, Trainspotter, Trannyboi, Transcendence, Trasman, Trav230, Travelbird, Traversc, Travis Cleveland, TreXiaNos, Tree Biting Conspiracy, Tregoweth, Treisijs, Trevdna, Trevor MacInnis, TreyHarris, Tribulation725, Triggerhippie4, Trinexx56, Trip Johnson, Tripacer99, Trongphu, Trontonian, Trottsky, Trovatore, True 505, Truecoder777, Truta1, Truthanado, Truthy not fakie, Ttwaring, Tubular, Tuckerdaman, Tud123, Turel, Turgeis, Turian,

Vacation, PAWiki, PCHS-NJROTC, PKM, Padvi, Pandaslayer69, Papercrab, Parkwells, Parsecboy, Pascal, Pastricide, PatGallacher, Patar knight, Patrickneil, Paulmcdonald, Pdakeyboardplaya, Pearle, Peetlesnumber1, Pekinpekin, Penguinluvr22888, Penndot1, PennsylvaniaGov, PeterSymonds, Peteritism, Pgan002, Pgk, Phantom784, Phantomsteve, Pharaoh of the Wizards, Phgao, PhiladelphiaCVB, Philip Trueman, Philippe, Phillies1fan777, Phillip J, PhillyPartTwo, Piano non troppo, Picnicbystream, Pigman, Pikeweatherman, Piledhigheranddeeper, Pilotguy, Pilover819, Pinethicket, Pitt99, Piuslxix, Planckton, Pne, Pnkrockr, Poccil, Poeloq, Polaron, Polarscribe, Pollinator, Postdlf, Praline97, Prestonmag, Ps0313, Psy guy, Purpleblack123, Pwh4130, Pwt898, Pwu2005, Pzavon, Qatter, Qqqqqq, Quantpole, Question12, Quilokos, Qwertyus, Qwyrxian, Qxz, R'n'B, R3D3Y3S27, RBBrittain, RJN, RJaguar3, RMHED, RNLion, RUL3R, Radon210, RaggedBiscuit, Rah126, Ram-Man, RamBow, Ranieri66, Ravedave, Ray Glock-Grueneich, RayBirks, Rayoflight278, Razorflame, Reagank, Real decimic, Realm of Shadows, Recognizance, Red Winged Duck, RedHillian, RedShiftPA, RedWolf, Redneb, Reedy, Retired username, Rettetast, Revas, Reywas92, Rfc1394, Rfl, Rgrof, Rhashan1, Rhatsa26X, Rich Farmbrough, RichardF, Rick Block, Ricky81682, Rickyburger15, RingtailedFox, Rjd0060, Rjwilmsi, Rmhermen, Rmosler2100, RobertG, Robhmac, Romanm, Ronark, Ronhjones, Ronnocian159753, Roodog2k, Root Beers, Rorygibbensrolph, Rossdegenstein, Rpfitzgerald, Rreagan007, Rrius, Rsherman40, Rsieg, Rufus843, Ruhrfisch, Ruok9c, Rwboa22, Ryan wallack, Ryulong, S-1-5-7, SB Johnny, SNIyer12, STEPHENFINAFROCK, Sad gurl, Salvioshia, Sam Staton, Sammy Houston, Samuel Blanning, Sanbeg, Sango123, Sannse, Santa-411, Saopaulo1, SatyrTN, Saurarity, Savatar, Saz127, Sbressi, Scanlan, Scientizzle, Scjessey, Scope creep, Scorpio1981, Scranton215, Scsmatt, Sean.hoyland, Search4Lancer, Seb az86556, SebastianHelm, Sfan00 IMG, Sfmontyo, Sgoldman10, Shanes, Shark96z, Shaunv123, Shiglhil, Shiquasa, Shoaler, Shouta, Shrommer, Shuffdog, Sietse Snel, Sillyserhan, Silvermoonx, SimonP, Skabat169, Skarebo, Skaterfunk, SkerHawx, Skiasaurus, Skizzik, Skunkboy74, Sky Attacker, Slambo, Slipperyweasel, Sloman, Sloshmo, Slushy9, Slvrstn, Smarkflea, Smith03, Snellmann, Snow Blizzard, Snoyes, SoCalSuperEagle, Socially Isolated Hamster, Solipsist, Some jerk on the Internet, SomeFreakOnTheInternet, SomeUsr, Sonarclawz, Sonicobbsessed, Sourpatchkids666, South Bay, South Philly, Sp33dyphil, Sparkhurst17, Spence737, Spiesr, Spikebrennan, Spitfire, SpookyMulder, Spotteddogsdotorg, Sprigdog, Spyder Monkey, Spydrlink, Spyneyes, Sriharsh1234, Sroulik, Sshikora, StAnselm, Standejb, StaticGull, Stbalbach, Steel, SteelMariner, Steelcityartst, SteffenB, Steinsky, Steve03Mills, SteveSims, Strannik, Strongbad1982, Suffusion of Yellow, Sullivan9211, Sunborn, SuperHamster, Superbeecat, Svgalbertian, Swanseamom, Swarm, SweetNeo85, Swollib, Sysm0921, T1980, TAnthony, TPIRFanSteve, TUF-KAT, Tangotango, TastyPoutine, Tbhotch, Tcncv, Tdmthomas, TecmoBo, Tedernst, Tedickey, Telekenesis, Template namespace initialisation script, Tesscass, Tetraedycal, Tgeairn, The Epopt, The Evil Spartan, The High Fin Sperm Whale, The Man in Question, The Thing That Should Not Be, The Transhumanist, The Utahraptor, The Wednesday Island, TheAlphaWolf, TheDJ, TheHoosierState89, TheRanger, Thebob000, Thegish, Thegreatdr, Thekrazykerzman, Thenameswill, Thesouthernhistorian45, Thetrollerstroll, Thingg, ThisguyYEA, Thisisbossi, Thomas Paine1776, Thunderbunny, Tide rolls, Tidying Up, Tiller54, TimBentley, Timcough, Timmyxcore666, Timneu22, Tinkey2, Tiptoety, Titoxd, Tiyang, Tlesher, Tlonca, Tnxman307, Tobby72, Tohd8BohaithuGh1, Tom Morris, TomStar81, Tomdobb, Tommy2010, TommyBoy, Tompie913, Tomtom, Toniokiv, Tony1, Toobevr1244, Toohool, TortoiseWrath, Towsonu2003, Tpbradbury, Tracer9999, Trafford09, Traforad, Tranquility2007, Travellingstone, Tresiden, Trimalchio, Triscuitface, Trugster, Trusilver, Truthanado, Tstrobaugh, Tunafizzle, Tuneguru, TwinCityIL, TwinsMetsFan, Twobestfriendz, Tyrol5, Ufwuct, Ukexpat, Ulmanor, Ulric1313, Ultimadesigns, Unity-Girl, Unyoyega, UsefulWikipedia, User5437, Username is shelby, Ussav, Uzusan, VB101, Vafthrudnir, ValleyRoad19046, Vanished user 194difuh2ruhqwdoinxojakdjncno234r, Vanished user ewfisn2348tui2f8n2fio2utjfeoi210r39jf, Vanished188, Vasiliymeshko, Vegaswikian, Veinor, Velvetron, Verrai, Versus22, Violinobaixo, Viridian, Volphy, Vpuliva, Vrenator, Vsmith, Vzbs34, WLRoss, WadeSimMiser, Wafulz, Wapcaplet, Wars, WashingtonWillie, Wavelength, Wayne Slam, Weatherman1126, Welsh, WereSpielChequers, Wereon, West Brom 4ever, Wester, Whatcanbrowndo, WhisperToMe, WikHead, Wiki Mateo, Wiki alf, WikiZorro, Wikibofh, Wikichanger22, Wikigirl98, Wikiwiji, Wild Wolf, Williamr10, Willking1979, Willydick, Wilsongirl00, Wimt, Wissgrl95, Witchzilla, Wknight94, Wmahan, Woohookitty, Woudloper, Wrightchr, Write On 1983, Wtmitchell, Wyattrahl, X292FirefighterX, Xb-70, Xfatmoe, Xyzzyva, YUL89YYZ, Yamamoto Ichiro, YankeeDoodle14, Yanksox, Yanwen, Yappy2bhere, Yarvin, Yayme23, YellowMonkey, Yilloslime, Yo its yamom, YodaRULZ, Yopure, Youngamerican, Your Mom3740, Yuenglingmaster, Zarcadia, Zavtrakat, Zeno22, Zereshk, Zimbabweed, Zoe, Zoicon5, Zokashi, Zooando, Zulitz, Zundark, Zzedar, Zzyzx11, Érico Júnior Wouters, პაატა შ, , 3001 anonymous edits

**Lebanon,_Pennsylvania** *Source*: http://en.wikipedia.org/w/index.php?title=Lebanon%2C_Pennsylvania *Contributors*: 24rhhtr7, Acntx, Ahoerstemeier, BMRR, Badagnani, Badbilltucker, Bbert1994cchs, Binksternet, Bob Burkhardt, BokicaK, Cameron8782, ClockworkLunch, Comoflebassoc, Cramigo, Cwhoffman1, DMacks, Dogface1, Ed g2s, Ermanon, FF2010, Falcnsoccer10, Gerry D, Gilliam, Gogo Rulez, Goodguitar, Heff01, Heyhey Hey, Hodgepodgeusa, Hooahman, Ian Pitchford, JRob, Jermerc, Jesselenton, Jklin, Jllm06, John Cardinal, John K, Joshdick, Jroth.x, JustinTime55, King of Hearts, Kingturtle, KnoxSGT, Koryakov Yuri, Law1260, Liz2330, MJCdetroit, Mareino, MarnetteD, Mayumashu, McSly, Mjg9461, Moonpuppy9, Moriori, Nhartman, Nojika, Nuge, Nyttend, Nyutko, Obird25, OverlordQ, Pearle, Peetlesnumber1, Pha telegrapher, Pigsonthewing, ProjectStimpy, Pubdog, PurpleChez, Quartermaster, RA13841401, Rafael M Torres, Ram-Man, ReadQT, Rjwilmsi, Robertjohnsonrj, Schzmo, Scott Shay, Seaphoto, Search4Lancer, Sensor, Shanes, Skabat169, Sleddog116, Solar-Wind, Stephenchou0722, Stevgee, TarenCapel, TehDrew, TerminalPreppie, Thisisbossi, Tinton5, TooPotato, VerruckteDan, Whhalbert, Wildnox, WillC, Wilsongirl00, Winchelsea, Wrightchr, Xnatedawgx, 173 anonymous edits

**United_States_Census_Bureau** *Source*: http://en.wikipedia.org/w/index.php?title=United_States_Census_Bureau *Contributors*: 007bond, Abductive, Acsenray, Aephyx, Ahoerstemeier, Al Lemos, Alan d, Alex1011, Alexandria, AndrusTandrus, Anna Frodesiak, AnonMoos, Antandrus, Armonica, Astrocommander, Asturianu, Atlien, Aude, AutoGeek, Babij, Beland, Bender235, Big Adamsky, Bikepunk, Bill.lay, Biruitorul, Bkonrad, Blargh29, Bonder00700, BradVanDam, Brandon5485, Brookie, Buaidh, Buster7, CMG, CambridgeBayWeather, Can't sleep, clown will eat me, Cancandance, CapitalR, Cartoonmoney, Chamal N, Chemboss, Chrism, Ckatz, Clayoquot, Clindberg, Closeapple, Cocircular, Coffee, Colonist, Compfreak7, Conversion script, CopperSquare, CsDix, Csyberblue, Curps, Cybercobra, Darkfight, Daven200520, David Kernow, Davilaa, Davodd, Den fjättrade ankan, Deodar, Der Träumer (DE), Dipankan001, Discospinster, Dmgaldi, Dough4872, Download, Dr.PeterRobertson, Dthomsen8, Duffman, EOBeav, ESkog, Eastlaw, Econr6, Eeaee, Elakhna, Epbr123, Epolk, EurekaLott, Evercat, Farquaadhnchmn, Favonian, Formulax, Framed0000, Frank101, Frazzydee, Furado, Fæ, GABaker, Gadickson, Gauth001, Geraldo Perez, Gidonb, Giftlite, Globalsolidarity, Gnominonwiki, GoldRingChip, Gpollock, Grafen, GregAsche, Ground Zero, Gugganij, Happenstancial, Harro5, Haruo, Hello71, Hephaestos, Hoppingalong, ImperfectlyInformed, Izzy007 1, JONJONAUG, Jengod, Jerryseinfeld, Jholman, Jim1138, Joao Xavier, Joseph Solis in Australia, Jusdafax, Jvcdude, Kbdank71, Ken Gallager, KenFehling, Klaam, Koavf, Kochamanita, Krellis, LOL, Lear's Fool, Leif, Leinad, Leszek Jańczuk, Levineps, LizardJr8, Lotje, Luna Santin, Magioladitis, Markjohnson87, Marshman, Martinwj53, Materialscientist, Mav, Meaghan, Mediran, Megaman en m, Michael Hardy, Microtony, Minesweeper, Musichman, Mxn, MyTwoCensus, Mysid, NickCT, NicoNet, Nightscream, NoSeptember, O18, Ohnoitsjamie, Olivier, One, Parkwells, Patrick, Pax:Vobiscum, Peak, Pearle, Pi, Pi zero, Pittimann DE, Pittimannl, Polmandc, Postdlf, Reach Out to the Truth, ReadQT, Rettetast, Revth, Ricav, Rich Farmbrough, RingtailedFox, Rklear, Robert Merkel, Rougher07, Rwwww, Sardanaphalus, Sarranduin, Scriberius, Semolo75, Seth Goldin, Shannon bohle, SimonP, Sk8tercolt18, So-nj, Spangineer, Srikeit, Srushe, Ssbohio, Stelmakia, Storm Rider, TJMcDJ, TOttenville8, Tanweer Morshed, Tedickey, Tenthfirst, Textorus, The Anonymouse, The Epopt, The Nut, Themfromspace, ThinkingTwice, Tide rolls, Tim!, Tisane, Tom, Tootallmaan, Touchmytralala, TygerDawn, UH Collegian, Vanka5, Ventur, Vuurcifer, Wars, Wayne Slam, West.andrew.g, WhisperToMe, Widr, WikiEditor5493, Willking1979, Yngvadottir, Yone, ZeWrestler, Zzuuzz, شرز , 231 anonymous edits

**Census** *Source*: http://en.wikipedia.org/w/index.php?title=Census *Contributors*: 16@r, 2toise, 90 Auto, A. B., ASDFGH, Abhaga, Achim Jäger, Adam Bishop, Aewold, AgainErick, Ahmedrefat, Ahmet Huner, Ahuskay, Aitias, Alansohn, Alex.muller, Alexd321, Alexzr88, Allstarecho, Altenmann, Andres68, Andrewpmk, Andy Marchbanks, Andycjp, Angrytoast, Angusmclellan, AnonMoos, Antiuser, Arcadian, Argon233, Arkaig1, Arnab1996, Artaxiad, Ashley thomas80, Australia boy, Avala, Avenged Eightfold, Avoided, BD2412, Badanedwa, BanyanTree, Barefeetdude, Barkers PR, Bart Versieck, Bender235, Benlisquare, Berean Hunter, Berkunt, Bersus, Bffjil, Bfswebstat, Bhadani, Bigsean0300, Birnuson, Biruitorul, Bitbut, Blackknight12, Blehimbored:), Bloodshedder, Bobblewik, Bogdangiusca, Bongwarrior, Boomer1234567, Boreas74, Boxplot, Brandmeister (old), Brianjd, Brutalhovno, Bry9000, Bucketsofg, CRICKETLtd, CTF83!, Ca3tki, Calliopejen1, Calvintiger, CanOfWorms, Canadaonline, Carboli, CaribDigita, Carwil, Cdc, Celtmist, Cgingold, Cherkash, Chester Markel, Chirag, ChrisGualtieri, ChrisNickson, Chrislk02, Chrism, Christian List, Chrono1084, Chuq, Ckatz, Closedmouth, Cmapm, Cnees, Colchicum, CommonsDelinker, Conversion script, Coopkev2, Cordless Larry, Courcelles, Cpl Syx, Crazyquesadilla, Cribbswh, Cstreet, Cup o' Java, Curb Chain, Cyanoa Crylate, Cymru.lass, D. F. Schmidt, DBigXray, DS1953, Dale Arnett, Dan653, DanielCD, Datu Dong, David Thrale, David1217, Davidcannon, Dbachmann, Dcastor, Ddroar, Deeptrivia, Delirium, Denisarona, Dewet, Dgkuty, Die spam die, Dinakarr, Dirkbb, Discospinster, Doggitydogs, Dominictroc, Donald Albury, Donfbreed, Dongjing, Doric Loon, Doseiai2, Dramatic, Drozmight, Dub8lad1, Dwight666, EamonnPKeane, Earl Andrew, Edcolins, Edward, Eilthireach, Eleassar, EmanWilm, Emilios1310, Emperorbma, EqualRights, Eregli bob, Erianna, Eric Wester, Erkan Yilmaz, Estoy Aquí, EugeneZelenko, Evil Monkey, Exec8, Ezhiki, Facial, FakTNeviM, Farhan Zafar, Fatstraw, Favonian, FedLawyer, Fern Forest, Fg2, Flapdragon, Flying Saucer, FocalPoint, Footballover, Fratrep, Frazzle, Fredbauder, Freshbakedpie, Futurist110, Fz22, Fæ, GABaker, Gaius Cornelius, Galoubet, Garrettmca, Gazilion, GenericBob, Genev82, Geoffrie, Gidonb, Giraffedata, Gogo Dodo, Gointemm, Gookhatingvet, Gordon.lunar, GraemeL, Graham87, Granucci, Grick, GringoInChile, Ground Zero, Guest9999, Gun Powder Ma, H2g2bob, H@r@ld, HJ Mitchell, Hadal, Hagedis, Hairy Dude, Halibutt, Hank7hill, Hattrem, HenryLi, Hephaestos, Heyhey Hey, Hibernian, Hmains, Hohum, Homerjay, Hut 8.5, Hwy43, Hydrogen Iodide, IRP, IRWolfie-, Igoldste, Ihatepotsmokinghippies, Iiiiaaaa, IkonicDeath, Imoen, IndepIntel, Intgr, Iridescent, Isaac120, Isthmus, Itsmejudith, JForget, JHunterJ, JackofOz, Jagged 85, Jannismakris, Jauerback, JavierMC, Jboyle4eva, Jeandré du Toit, Jeff G., Jeraphine Gryphon, Jerry Ritcey, Jerzy, Jibril23, Jim Huggins, Jim no.6, Jimp, Jiri007, Jm1122k, Jnestorius, Joao Xavier, JoeSmack, John ellenberger, John of Reading, JohnGurney, Johnjay1745, JonHarder, Jonemerson, Jonesey67, JorgeGG, JorgeRodriguez, Joseph Solis in Australia, Joseph simeon, Josette, Julius Sahara, Julo, Jusdafax, Jusjih, Just plain Bill, Just4ikarus, KGasso, Kaihsu, Kat263, Kay556690, Kayau, Keith Edkins, Kembangraps, Khalil babli, Kiefer.Wolfowitz, Kigali1, Kim bas, King Lopez, KingKane, Kingpin13, KirrVlad, Kku, Koroner, Krikwen, Kristaga, Kross, Kyky108, L Kensington, Lacrimosus, Laogeodritt, Lasunncty, Lateg, Lauzonda, Lawlessurchin, Ldsviking, Legumeman, Leszek Jańczuk, Levineps, Li-sung, Licor, Lindar, LizardJr8, LlywelynII, Llywrch, Logicwiki, Ltool63, Luna Santin, Luntertun, LynxNukem, Mack2, MadGeographer, Magbatz, Malcolm Farmer, Mandarax, Mandsford, Manuel Anastácio, Marathone, Marek69, Mark Arsten, Martpol, Masoudfcr, Materialscientist, Matt Brennen, Matthead, Mattsains, McSly, Meiskam, Meister, Melcombe, Mentifisto, Mhym, Michael Devore, Michael Hardy, MichaelCarl, Midway, Mikkel, Milboz, Mild Bill Hiccup, MisfitToys, Misskaykay, MithrandirAgain, Mkanoni, Modulatum, Mr.Badlands, MrBell, MrFawwaz, Mrh30, Muppet, Mxn, MysticalCow, Mzajac, Mátyás, Nabob Nagusa, Nachoman-au, Nateofgreat, Nauseam, NawlinWiki, NeonMerlin, NerdyScienceDude, Nestled like spoons, Neurolysis, Newhoggy, Nfitz, Nhgenealogy, Ninsheart, Niteowlneils, Nmacu, Nn123645, Nomi887, Norbu19, Notheruser, Nponte, Nrlight, Ns.code, NuclearWarfare, Numbo3, Nunquam Dormio, Nyttend, O'Dea, Ohconfucius, Oleg Alexandrov, Oliver Pereira, Olyus, Omdfw, One, Onya4ever, Optimist on the run, Orodreth, PANONIAN, PMLF, PRODUCER, Panapp, PaoloValente, Parishan, PatrickFlaherty, Paul Norfolk Dumpling, Paul W, Paul-L, Pepith, Pgallert, Phantomsteve, Pharaoh of the Wizards, Philip Trueman, Philopp, Pigman, Pinethicket, Piotrus, Plastikspork, Pne, PoeticVerse, Polly Math, Poor Yorick, Priyansh verma, Prop3v56, Quadell, Quadparty, Quebec99, Queenmomcat, Quicksilvre, Qwertzy, Qwfp, RODERICKMOLASAR, RTG, Rami R, Rammy2cool, Razor2988, Reach Out to the Truth, Reaper Eternal, Red Slash, Reinyday, Remigiu, Ren Sydrick, RenamedUser01302013, Rentzepopoulos, Rhopkins8, Rich Farmbrough, Richard Arthur Norton (1958- ), Rjwilmsi, Rmhermen, Rollon, Romanm, Ronhjones, Rui Gabriel Correia, Ryl411, Saddhiyama, Sahasrahla, Sandstein, Sannse, SatyrTN, Savidan, Scientus, ScottyBerg, Sdaeun, Seaphoto, Selket, SemperBlotto, Senator2029, Sesu Prime, Sfs90, Shadowjams, Shannon bohle, Shii, Shoessss, Siamax, Sift&Winnow, Simetrical, Sionk, Sivuman, Sjakkalle, Skizzik, SlimshadY, Smallman12q, Smarkflea, Smooth O, So-nj, Sombrero, Sortcoach, SpacemanSpork, Spearhead, Specs112, Spoladore, Srich32977, StAnselm, StaticGull, Stet, Sukarnobhumibol, Tarih, Tarun mor50002, Taztouz, Tempac, Terovian, The Rambling Man, The Thing That Should Not Be, TheRingess, Thunderbuck ram, Tide rolls, Tim1357, Timc, Timneu22, Timrollpickering, Tomcakes, Tomi, Tommy2010, Tomsega, Tony Corsini,

Tpbradbury, Treisijs, Triona, Truthanado, Tualha, Twhyatt, Typing General, UberScienceNerd, Ultraboy8888, Uncle Dick, Unnikn, Urg writer, Utar, Valhalla, Vanished user 342562, Vicki Rosenzweig, VictorianMutant, Wavelength, Wayne Riddock, Wheatleya, WhiteWriter, Whodhellknew, Wienertakesall, Wiki-uk, Wiki1609, WikiPuppies, Wikidan829, Will Beback Auto, Wimt, Wolfvuk, Wtshymanski, Wysprgr2005, XNanakate, Xanthar, Ycu1101ycu, Yidisheryid, Yngvadottir, Ynhockey, Yom, Yoman82, YuMaNuMa, Yupik, Zaian, Zangar, Zigger, Zinc2005, Zlerman, Zollerriia, Zundark, Zxcvn80, Zzuuzz, ن‌گون‌ابن‌گون‌ی, اشف ع‌قیل, 847 anonymous edits

**Population_density** *Source*: http://en.wikipedia.org/w/index.php?title=Population_density *Contributors*: 1234r00t, 1297, 21655, 2mcm, 45Factoid44, 4halsop, A-giau, A. Parrot, A.Ou, Abhijitsathe, Achim Jäger, Addihockey10, Agentbla, Ahoerstemeier, Alansohn, AlexD, Alexjohnc3, Andre Engels, Andycjp, Anthonyz747, Aphaia, Apokryltaros, Applecharm, Aranel, Areology, Astuishin, Aude, Avicennasis, Aziz1005, B.d.mills, Barend, Barneca, Biederman, Blacklite, Blaircorrigan, Blanchardb, Bobblewik, Bogdangiusca, Br77rino, BrainyBabe, Brianlo747, Bricklayer, Brisvegas, BrokenSegue, Bsadowski1, Buchanan-Hermit, Bwilkins, CUSENZA Mario, Calabe1992, Candorwien, Catgut, Cdc, Cenarium, Chimpex, Ckatz, Cliff smith, Clovis Sangrail, Coffee, CommonsDelinker, Corpx, Courcelles, Cretog8, Cwgordon7, D. Cerv, DCEdwards1966, DL81292, DancingPenguin, Darth Panda, Davidcannon, Dbfirs, Dekisugi, Desiphral, Diaboli, Docu, Donarreiskoffer, DoorFrame, Dragoon17, Drrngrvy, Dusti, Dysprosia, E Pluribus Anthony, ERobson, Eagle4000, EdGl, Edgar181, Edward, Eeyore tim, Egg Centric, Eggman64, Eleassar777, Elekhh, Elmsat, Elockid, Eopsid, Epbr123, Eric Sandholm, EugeneZelenko, Excirial, Extra999, Eyedubya, Eyeofhorace, Farosdaughter, Figureskatingfan, Filemon, Francium12, Frozen4322, Func, GFW82, Garion96, Gcorral, GeeJo, Geremia, Gergis, Ghaly, Gidonb, Ginsengbomb, Glen, Gnosis1185, Gog, Gparalikidis, Graculus, Grafen, GrayFullbuster, Gryffindor, Gwabell, HGB, HalfShadow, Hallows AG, Hamedog, Helix pomatia, HenryLi, Hermitage17, Heron, Hitman012, Hoplophile, HornetMike, Hsm rocks ya right, Hypertall, Hégésippe Cormier, INeedANameThatIsFunny, Ibbn, Introvert, Iridescent, Island, J.delanoy, JNW, JUSTM3-X, Jacob.jose, Jake Spooky, Jakeburberry, JavierMC, Jaw959, Jdeamerr, Jds10912, Jeff G., Jim457, Joblio, JohnDoe0007, Johnuniq, Jonkerz, Jorgeposada14, Joseph Philipsson, Joseph Solis in Australia, Josh howson, Josisb, Jpk, Jprksh, Jrein59, Kaare, Karl-Henner, Karok, Keenan Pepper, Keilana, KenFehling, Ketiltrout, Kevin Gorman, Klparrot, KnowledgeOfSelf, Kr4ft, Kransky, Kurieeto, Kyle1278, L Kensington, Lakers, Latebird, LcawteHuggle, LeaveSleaves, Leks81, Leszek Jańczuk, Levineps, Liftarn, Lightmouse, Logose, Lotu, Lucas175, Lugia2453, LukeStewart, Luxborealis, Lytsio, MER-C, MJCdetroit, MaNeMeBasat, Mack2, Malcolmxl5, MapsMan, Marc Lacoste, Marek69, Mark Arsten, Mark J, Martial75, Materialscientist, Matthewcgirling, Maximus Rex, McSly, Melcombe, Mgnbar, MidgleyDJ, Mightymights, Miguel.mateo, Mindmatrix, Mintleaf, Missmarple, Moreschi, Motoxrampage101, Mrbuddhafreak, Mschlindwein, Mxn, NLikhov, NTK, Nagy, Nakon, Naval Scene, Neko-chan, Nentrex, Nepenthes, NewEnglandYankee, Newbyguesses, Nick C, Noah Salzman, Nunquam Dormio, Nyttend, Olly150, Olsonnnnn, Ongar the World-Weary, Oreo Priest, Orestimba49, PBP, Panicburststs, Patrick, Pax:Vobiscum, Pearle, Pengo, Pgan002, Phaust, PhilKnight, Philip Trueman, Physchim62, Piano non troppo, PierreAbbat, Pigman, Pinethicket, Piotrus, Pleasetouchme, Popsracer, Professionalsmartpeople1, Pumpmeup, Quasipalm, R Lowry, R'n'B, RLeopard, Radishes, Readyhairman, Red Slash, Remuel, Rev. John, ULC, RexNL, Rgdboer, Riana, Richard Warrington, Rmhermen, Romanm, Ronhjones, RoyBoy, Rushbugled13, Ryaelt, Ryan Valdés, SDC, SFK2, Samrager, SarekOfVulcan, Sarenne, Scohoust, Shizhao, Simmania, Simonkeuler.1245, Sin Harvest, Skinnyweed, SmthManly, Snowolf, Snowolfd4, Sonicyouth86, Spencer, Srasku, StaticGull, Ste, Steven Zhang, Steveprutz, Stormie, Student7, Sumahoy, Sunoco, Superm401, Surfbear335, Symode09, TGalt, TUKAKKAR, Tabitcome2, TaintedMustard, TakuyaMurata, Tangotango, Tanzy97, Taykp, Tbhotch, Tellyaddict, Terence, The Angel of Islington, The Anome, The Determinator, The Hybrid, The Thing That Should Not Be, Thecaeser, Thehelpfulone, Thekohser, Theopolisme, Thermalite, Thucydides of Thrace, Thunder8, Tide rolls, Timeshifter, Tktktk, Tlc61, Tobias Galtieri, Tom harrison, Tomatoman180, Topbanana, Torchwoodwho, Tradewater, Tree Kittens, Trengarasu, Triesault, Tsxado, Tuanglen, Una Smith, Uncle Dick, Uuuussssseeeerrrrnnnnaaaammmmeeee, Vegard20, Violetriga, Vollex, Vranak, Waggers, Waldir, Wangi, Wasbeer, Wavelength, Waywardhorizons, WeißNix, Wereon, Will Beback, Wilsbadkarma, Wimt, Yellowtruck, Ylem, Yngvadottir, Zanimum, Zap Rowsdower, Zfr, Zzuuzz, Александр, आशिष भटनागर, 718 anonymous edits

**Time_zone** *Source*: http://en.wikipedia.org/w/index.php?title=Time_zone *Contributors*: *drew, 159753, A. B., A.Ou, A0404141041, A8UDI, AHeneen, ANDREJACK, Aaaallleex, Aaronw, Absalom89, Acegikmo1, Acs4b, AfDproXX, Ahoerstemeier, Alanmak, Alansohn, Albmont, Alex Bakharev, Alksentrs, AllyUnion, Alynna Kasmira, Ambarish, Aminullah, Amit6, Amolkold, Amorim Parga, Amorymeltzer, Andre Engels, Andrewpmk, Androl, Anneman, Antandrus, Arichnad, Arthena, Ashley Y, Asterion, Australia boy, AxelBoldt, BD2412, BIL, Banhtrung1, Basil.bourque, Bastin, Bazonka, Bdodo1992, Bearcat, Beland, Bellczar, Ben Arnold, Bentogoa, Bermlee, Bethpage89, Big Adamsky, Billybob112, Bjankuloski06en, Bjarki S, Bjcairns, Blahma, Bluelion, Bobo192, Bomac, Bongwarrior, Breno, Brian Kendig, Brian Patrie, Bryan Derksen, Budlight13, Buggery Boar, Bunthorne, Bushcarrot, CQ, Caese, Calor, Calton, CambridgeBayWeather, Can't sleep, clown will eat me, Canjth, CaribDigita, CatherineMunro, Celarnor, Cenarium, Centpacrr, CesarB, Charlene.fic, Chihuahua State, Chipmunkdavis, Chmod007, Chochopk, Chris24, ChrisGualtieri, Christy747, Chuck SMITH, Chun-hian, Chuq, Chuunen Baka, Circeus, Cirt, Classical geographer, Cliffb, Closedmouth, Cnmuk, Coder Dan, ColinKennedy, Colonel Cow, Commdor, CommonsDelinker, Comrade Graham, Confer44, Conscious, Conversion script, CopperSquare, Corumplex, Crd721, Cst17, Curps, Cyrus the virus grissimo, D-Rock, D15724C710N, DARTH SIDIOUS 2, DMarvin, DO'Neil, DPH1110, DVD R W, Da Vynci, Dale Arnett, Damian Yerrick, Damirgraffiti, Darth Panda, David Biddulph, DavidGrayson, Davidcannon, Davodd, Dcsohl, Deelkar, Denelson83, Derek Ross, Derild4921, Disill, Diti, Djszucs, Dmacgr 22, Dmeranda, Dmh, Dmn, DocWatson42, Docu, Donarreiskoffer, Doniago, Donner60, DoubleBlue, DouglasCalvert, Download, DrBizarro, Dragon taker, Drbreznjev, Drbug, Dreadstar, Drh104, Dsyzdek, Duncan, Duncan7670, DyNama, Dysepsion, E. Ripley, ESkog, EamonnPKeane, Earl Andrew, Earthlyreason, Ebelular, Ed Poor, Edward, Eequor, Eerlend, Ehrenkater, Ekren, Enchanter, Eomund, Eoujw, Epbr123, Eraserhead1, Erasmussen, Eric Herriman, Eric-Wester, Eric119, Esmito, Esperant, Eubulides, Euchiasmus, Evice, Evil Monkey, Evil saltine, Ewawer, Excirial, Ezhiki, Ezhuttukari, F. F. Fjodor, FB 2000, FF2010, Facts707, Fanyavizuri, Farosdaughter, Farzanegan, FayssalF, Federico Tortorelli, Feedmecereal, Feezo, Feinoha, Fennec, Ferkelparade, Fffdervis, Fieldday-sunday, Fmph, Fonadier, Forseti, FrancoGG, Freddiecaveman, FreplySpang, Fresheneesz, Funandtrvl, Fundamental metric tensor, Fuzheado, Gabbe, Ganteng, Garfield226, Gary King, Gerry Ashton, Giant89, Giftlite, Glane23, Glenn, Glenn L, Gogo Dodo, Golbez, Goobergunch, GorillaWarfare, Gorrister, Goudzovski, Graeme Bartlett, Grafen, Graham87, Grassfire, Grick, Groogle, Ground Zero, Grstain, Guaka, Gzornenplatz, H2g2bob, HJ Mitchell, Hadal, Hadiyazzi, Haham hanuka, Haihe, Haikupoet, Hajor, Happening1, Hapsiainen, Harp, Harriv, Hazhk, Heb, Heitordp, Hellerick, Henrygb, Hesperian, Hi is bob, HistoryAddict, HortusBlizzard, Htmlland, Humilulo, HumphreyW, Hvn0413, Hydrargyrum, IKidaround, IMSoP, IQveteran, Ido50, Ignis Fatuus, IgorMagic, Imoen, Imz, Incnis Mrsi, Indefatigable, Iner22, IngSoc BigBrother, Instantnood, Intgr, Iridescent, Island, Itai, Itisonly4me, Ixfd64, J-stan, J.delanoy, J.smith, JHunterJ, JLaTondre, JNW, JackofOz, Jakes18, Jamadagni, Jarrodiscool141314, JavierMC, Jc3s5h, Jcra, Jd4508, Jda, Jeff G., Jeff3000, Jeffq, Jerome Charles Potts, Jerzy, Jesper Norgaard, Jiang, Jim1138, Jimp, Jj137, Jla, Jnestorius, Joe Kress, JoeSmack, Joey2cool2007, John Paul Parks, John Quincy Adding Machine, John of Reading, JohnOwens, JohnWittle, Jojhutton, Jojit fb, JonathanJosephsen, Jonathunder, Joostjodel, Joseph Solis in Australia, Jpk, Jplimpz, JulDes, Juliancolton, Julianp, Jumbuck, Jusjih, Just H, Jérôme, Karl Palmen, Karl-Henner, KathrynLybarger, KayEss, Ke4roh, Keeves, KeithH, Kh2mohamed, King of Hearts, Kingal86, Kiwi128, Klemen Kocjancic, Knowledge Seeker, Krawunsel, Kubigula, Kukini, Kungfuadam, Kyle Barbour, Kylekieran, LOL, Lacrimosus, Lalah1294, Lambiam, Lambyte, Lamrongpeng5, Last Avenue, Lee J Haywood, LeeTraynor, Leeyc0, LeinsterLion, Leucius, Leuko, Lewisjon, LilHelpa, Lincmad, Lir, Lizzie Harrison, Llywrch, Logan, Lolm8, LorenzoB, Lotje, Lucamauri, Lugia2453, Luis Dantas, Luna Santin, M, MKar, Mac, MacTire02, Macaddct1984, Magnus Manske, Maksim L., Malrase, Mancunius, Marechal Ney, Marek69, Mariusmiti, Mark Leander, MarkSweep, Marshman, Martarius, Mathew12123434, MatthewVanitas, Mav, MaxDZ8, Mayfoev, Mayooresan, McSly, Mean as custard, Meelar, Meini, MelbourneStar, Mernen, Mert34, Meters, Michael Glass, Michael Hardy, MichaelBillington, Michfan2123, Midnightblueowl, Mild Bill Hiccup, Minipie8, Minna Sora no Shita, Miquonranger03, Miranda, Miros 0571, Mjs1991, Modulatum, Mojo-chan, Monkeyman, Monni95, Monoxide333, Montrealais, Moocha, MoraSique, Morga555, Mormegil, Mr Bartels, Mr. Ajax, MrWolfy, Mrk695, Mryg, Mwarren us, Mxn, N1ywb, Naive cynic, Nanshu, Natpowning, Naval Scene, NawlinWiki, Nealmcb, NellieBly, Neptune777, NewEnglandYankee, Nichalp, Nicholas Weiner, Nicolas Love, Nightstallion, Niteowlneils, Nitsansh, Nk, Northumbrian, Nortonius, Novalis, Numbo3, Numen, Nunh-huh, Octahedron80, Ohnoitsjamie, Oleksiy.n, Olivier, OpenFuture, Optimist on the run, OverlordQ, OwenX, Oxymoron83, P. S. F. Freitas, PDucks, Paddu, Pak21, Panos84, Paranoid, Pasquale, Patach, Patcs0013, Patrick, Paul August, Paul Koning, Paul Murray, Paullees, Pavel Vozenilek, Pejman47, Peregrine981, Persian Poet Gal, Peter Wisner, PeterThoeny, Petrig, Pgan002, Phearson, Philip Stevens, Philip Trueman, Phoenix B 1of3, Pi zero, Piano non troppo, Picapica, Picaroon, PierreAbbat, PimRijkee, Pm5, Pokrajac, Popsracer, Possum, Preslethe, Psb777, Psdekloe, Pseudonymous Rex, Pytom, Quadell, Quasipalm, Rabbit233224, Radiojon, RainbowOfLight, Randyc, Raymondwinn, Rchandra, Rclocher3, Reaper Eternal, Redrose64, Redwood57, Reisio, Renetus, RenniePet, Requestion, RexNL, Rhobite, Riana, Rich Farmbrough, Rjwilmsi, Rmallins, Rmhermen, Robsinden, Roland2, Romanm, Rory096, Royaume du Maroc, Rsrikanth05, Rt66lt, Russell E, S.Örvarr.S, SGBailey, SL93, SV Resolution, Sahgal, Samuel, Samw, Saros136, Sceptre, SchmuckyTheCat, Schneelocke, Schoen, Scipius, Scoelho86, Scott Roy Atwood, Scott3, Seba5618, Secfan, SeoMac, SephChicago, Sephiroth BCR, Sesu Prime, Seth Ilys, Sewing, Sheridan A.M. Hamilton, Shoefly, Sietse Snel, Simon J Kissane, Simon_J_Kissane, Simpsnut14, Siqbal, Sjorford, Skater, Skizzik, Skyman123, Skysmurf, Skywayman, Slawojarek, Smalljim, Smart77, Smithcraft36, Smjg, Smoken Flames, Snowdog, Sommerfeld, Sonninkikkeli, SouthernNights, Sp33dyphil, Spesh531, Spitfire, Squiffycat, Steel, Stephan Leeds, Stephen B Streater, Stephen.frede, SteveMX, Sunborn, Susurrus, Sverdrup, Swid, Swpb, T-resh, T.c.w7468, TGC55, TPK, TUF-KAT, TZ master, Tad Lincoln, Takita787, TakuyaMurata, Tarquin, Tashikelzang, Tbhotch, Tebloraf, Techman224, Techtonik, Teemu Leisti, Telempe, Tfts, That-Vela-Fella, The Duke of Waltham, The High Fin Sperm Whale, The Rambling Man, The Storm Surfer, The Thing That Should Not Be, The Tom, TheEgyptian, Thf1977, Thingg, Think outside the box, Thryduulf, Thumperward, Tide rolls, Tim Parenti, Time in Russia, TimeCurrency, TimeXCode, TimeZonesBoy, Timir2, Timneu22, Timwi, Tivolives, Tmangray, Tmol42, Tobias Conradi, Tomica, Topdeck, TopoChecker, Topory, Torpov, Transisto, TruPepitoM, TrueMahn, Tsja, Tutsuro, UhMaZe xpeK, Ultratomio, Uncle Scrooge, Undescribed, Uriber, Ute in DC, VampWillow, Vanished User 1004, Vanjagenije, Vaughan Pratt, Vee power, Velho, Versus22, Viriditas, Vsmith, Walshie79, Wangi, Watson24hour, Wavelength, WeißNix, Weregerbil, West.andrew.g, Wexcan, Widr, Wik, Wiki Wikardo, Wikidenizen, Wikilackey, Wikipediman, Will Beback Auto, Willking1979, Wissons, Wizard04, Woodstone, Woohookitty, Worldtimezone, Writ Keeper, Writherthebest100, Wrtiede, Wtmitchell, Wwutx, Xeryus, Ximi12, Xpi6, Xxeric thomasxx, Xxxx693, Xzqx, Y0u, YUL89YYZ, Yamara, Ybbor, YellowMonkey, Yosef1987, Yt788, Zakuragi, Zash20, Zidonuke, Zkzkz, Zoicon5, ZooFari, Zundark, Zzuuzz, Σ, Владимир Шеляпин, חלהק, 1556 anonymous edits

**Per_capita_income** *Source*: http://en.wikipedia.org/w/index.php?title=Per_capita_income *Contributors*: 0, 198.207.223.xxx, Accurizer, Aelius28, Ahoerstemeier, Ahuskay, Altaïr, Anir1uph, Antandrus, Arsenalfc06, Arthur Rubin, Ausir, Austinrok, Avenged Eightfold, BarretB, Bart133, Beland, Bhny, Black men in pants, Blue520, Bobo192, BrendelSignature, Brian0918, Brianhe, Brobdingag, Brockert, Bvu wieuwi, Calmer Waters, Canderson7, Carlosguitar, CarminPolitano, Cdc, Cheque some, Christian75, CieloEstrellado, Clearcontent, Cocoma, Curps, Cypoet, DARTH SIDIOUS 2, DFGGH, DJ Clayworth, DL5MDA, DO'Neil, DVD R W, Dalillama, Dark567, Dave6, DavidGC, Ddlamb, Deetdeet, Dekisugi, Denisarona, Driftwoodzebulin, Dtellett, DuRoad, Duoduoduo, Dustinkelley, Eastcote, Eastlaw, Elmondo21st, Elockid, EncMstr, Enviropearson, Epbr123, EurekaLott, Everyking, Evil saltine, Fingers-of-Pyrex, Foobar35, Fraggle81, Fram, Friosurely, GateKeeper, Gene.arboit, Giftlite, Girishsharma86, Gmaxwell, Graham87, Green Giant, Gtstricky, HIDECCHI001, Harlem Baker Hughes, Hgb1217, Ifnord, Ilikeverin, Imakegoodedits, Indon, Ixfd64, J.delanoy, JForget, JFreeman, JYolkowski, James086, JamieA, Jaranda, Jmarchn, Jogloran, John, Joseph Solis in Australia, Kkm010, LAX, Leslie Mateus, Leszek Jańczuk, Lfstevens, Longhair, Lotje, LuK3, Lunarune, MaCRoEco, Maurreen, Mike Payne, Miketwo, Morgan Leigh, Mxn, Myanw, Neechalkaran, Newkai, Nijusby, Nimur, Niteowlneils, Nuberger13, Pax:Vobiscum, Pdcook, Philip Trueman, Piano non troppo, Pinethicket, Polylepsis, Prashanthns, Priyatu, Quanneur, QuixoticKate, RachelleLin, Raymondwinn, Rebel, Reggionh, Resurgent insurgent, RexNL, Robtim86, Rocky Young Crobee, Rush Luagh, SLi, Shadowjams, Shanemcd, Simeondahl, SimonP, Snoyes, Spencerthomas55, Spiffy sperry, Sullykid77, Tagishsimon, TakuyaMurata, Tanvir Ahmmed, TastyPoutine, Taxman, Teapot207, Tekana, TheMidnighters, Thebestofall007, Thingg, Tim Ivorson, TimShell, Titodutta, Tobias Hoevekamp,

Toffee7053, Travelbird, Van helsing, Vary, VerruckteDan, Viridian, Volgar, Voyevoda, Walton One, Wekrjlsdlklksdrujl, Wiki alf, William Avery, Wimt, Wmahan, Zeppelin462, ناوقلاخ, 362 anonymous edits

**Poverty_threshold** *Source*: http://en.wikipedia.org/w/index.php?title=Poverty_threshold *Contributors*: Aaboltaabol, Acroterion, Aengus, AeriDenise, Agradman, Ahackworth, Aillema, Aleksd, Aliyevramin, All Is One, Allens, Amazingcreator79, Andy Xucha, Anomalocaris, Antandrus, Arctic Kangaroo, Arthena, Astronautics, AuburnPilot, Ausir, Barek, Beland, Bigpapi6292, Bluerasberry, Bookandcoffee, BrendelSignature, Burrburr, Byelf2007, CLW, Can't sleep, clown will eat me, Cantus, Cardsplayer4life, Cassivs, Cconnett, Cflm001, Chozanm, Chuckdevore, Cielomobile, Cirino73, Ckatz, Classicrockfan42, Closenplay, Cntras, Cocoaguy, Coelhodg, Commander Keane, Common Man, Cretog8, Cuvtixo, Dambrosio, Damicatz, DanKeshet, David Biddulph, DevilEditerLol, Dgies, Dismas, Dlohcierekim, Drmies, Duoduoduo, Dwrcan, Dystopos, Ed Poor, Edgar181, Edward, Egoadvocate, ElentariAchaea, ElinorD, Elitehobo, Epbr123, Eraserhead1, Eurosong, Fakelvis, Falconleaf, Fasttwitch, Fconaway, Fidalgo, Fitzaubrey, Foober, Frenzie, Gaff, Galf, Gareth Griffith-Jones, Gavin.perch, Gene.arboit, Gilliam, Gizziiusa, Globalsolidarity, Goldom, Grick, Guoguo12, Hadrianheugh, Hamedy1986, Haoie, HarveyA, Hector.C.Jorge, Hede2000, Henrygb, Heron, HexaChord, Hibernian, Hrh80, Human.v2.0, Humanflux, Hurutch, Ian13, Ivan1984, Iviney, J.delanoy, JakubATusaDotcom, Janshouse, JavaAndCSharp, Javadane, JeremyA, Jhortman, Jim1138, Jimtaip, Jitse Niesen, Joseph Solis in Australia, Josh Parris, Joy, Junling, Kanook, KathrynLybarger, Kellen, Kelly Martin, Ken@popp2.com, Khanra, Kirachinmoku, Kissmequick, KungFuMonkey, Lalit Jagannath, Lawrencekhoo, Leslie Mateus, Leszek Jańczuk, Levineps, LilHelpa, LilyKitty, Logical Fuzz, LolChavista, Lost tourist, MOA12345, Maralia, Marek69, Materialscientist, Mattbrundage, Matthewcgirling, Maurreen, Maxis ftw, McDutchie, Metta Bubble, Million Moments, Minimac, Misanthr0pe, Monsday, Mrshannigan, Mshecket, Muhandes, Musiphil, Mydogategodshat, Mynameisnotpj, Mysdaao, Möchtegern, Nach0king, Nagy, Naniwako, Naval Scene, Neptunius, Niceguyedc, NigelR, Nihiltres, Nishkid64, Noodleki, Noozgroop, Oceanflynn, Oledr, Otolemur crassicaudatus, P.bertazzi, Paddles, Passionless, Philip Trueman, Phoebe, Piotrus, Policy Person, Poopeater111, Postmortemjapan, Professortimithy, Quest for Truth, QuiteUnusual, R1000R1000, RJ010890, RainbowOfLight, ReiVaX, Revth, RexNL, Riana, Rich Farmbrough, Rjwilmsi, RobyWayne, Roke, SDnogo, Sam Hocevar, Sammo, Sapphic, Sarahj2107, Scarsxrunxdeep, SeanAhern, Search4Lancer, Septegram, Sh-otaibi, Shannon bohle, Sharkface217, Sharnak, Sheetal.sehgal, Shoy, SiobhanHansa, Skysmith, Slhurst, Smallman12q, Smokecrack, Snipson23, Squashy, Sreenand wiki, Staszek Lem, Strait, Tanaats, Team9a, Terjepetersen, The Mark of the Beast, The man stephen, Thingg, Tigga, Timc, Timwi, Tinton5, Toh, Tommy2010, TouristPhilosopher, Twang, Ultramarine, Valdomiro, Vanished user uih38riiw4hjlsd, Virtakuono, Volunteer Marek, Vungadu, WAGATHEE, Whatamldoing, Wing Nut, Woutersmet, Wtmitchell, Yaktx, Yoyoyoitsrob, Zappa711, Zouf, 378 anonymous edits

# Image Sources, Licenses and Contributors

**file:speakerlink-new.svg** *Source*: http://en.wikipedia.org/w/index.php?title=File:Speakerlink-new.svg *License*: unknown *Contributors*: User:Kelvinsong

**Image:wikisource-logo.svg** *Source*: http://en.wikipedia.org/w/index.php?title=File:Wikisource-logo.svg *License*: unknown *Contributors*: Guillom, INeverCry, Jarekt, MichaelMaggs, NielsF, Rei-artur, Rocket000

**Image:Map of Lebanon County Pennsylvania With Municipal and Township Labels.png** *Source*: http://en.wikipedia.org/w/index.php?title=File:Map_of_Lebanon_County_Pennsylvania_With_Municipal_and_Township_Labels.png *License*: unknown *Contributors*: US Census, Ruhrfisch

**Image:Map of Lebanon County Pennsylvania School Districts.png** *Source*: http://en.wikipedia.org/w/index.php?title=File:Map_of_Lebanon_County_Pennsylvania_School_Districts.png *License*: unknown *Contributors*: US Census, Ruhrfisch

**File:Flag of the United States.svg** *Source*: http://en.wikipedia.org/w/index.php?title=File:Flag_of_the_United_States.svg *License*: unknown *Contributors*: Anomie

**File:Great Seal of the United States (obverse).svg** *Source*: http://en.wikipedia.org/w/index.php?title=File:Great_Seal_of_the_United_States_(obverse).svg *License*: unknown *Contributors*: U.S. Government

**File:USA orthographic.svg** *Source*: http://en.wikipedia.org/w/index.php?title=File:USA_orthographic.svg *License*: unknown *Contributors*: User:Addicted04

**File:Increase2.svg** *Source*: http://en.wikipedia.org/w/index.php?title=File:Increase2.svg *License*: unknown *Contributors*: Sarang

**File:Indians giving a talk to Bouquet.jpg** *Source*: http://en.wikipedia.org/w/index.php?title=File:Indians_giving_a_talk_to_Bouquet.jpg *License*: unknown *Contributors*: Dantadd, Hsarrazin, Kevin Myers, Origamiemensch, Shakko, Uyvsdi, Wst

**File:The Mayflower Compact 1620 cph.3g07155.jpg** *Source*: http://en.wikipedia.org/w/index.php?title=File:The_Mayflower_Compact_1620_cph.3g07155.jpg *License*: unknown *Contributors*: Howcheng

**File:Treaty of Penn with Indians by Benjamin West.jpg** *Source*: http://en.wikipedia.org/w/index.php?title=File:Treaty_of_Penn_with_Indians_by_Benjamin_West.jpg *License*: unknown *Contributors*: Auntof6, Bukk, Frank C. Müller, Goldfritha, Hsarrazin, Man vyi, Nonenmac, Origamiemensch, Presidentman, RIbberlin, Shakko, Thuresson, Uyvsdi, Wmpearl, 3 anonymous edits

**Image:Declaration independence.jpg** *Source*: http://en.wikipedia.org/w/index.php?title=File:Declaration_independence.jpg *License*: unknown *Contributors*: Aavindraa, Amandajm, Americophile, Beyond My Ken, Bohème, Boo-Boo Baroo, Cambalachero, Editor at Large, Elemaki, Fred J, Geni, GrawpSock, Harpsichord246, Hluup, Ibn Battuta, Mhby87, Misogi, Morgan Riley, Mutter Erde, Nonenmac, Panoptik, Patstuart, Piledhigheranddeeper, Pmlineditor, Savh, UpstateNYer, WTCA, Wmpearl, Wst, Xavigivax, Zolo, 20 anonymous edits

**Image:U.S. Territorial Acquisitions.png** *Source*: http://en.wikipedia.org/w/index.php?title=File:U.S._Territorial_Acquisitions.png *License*: unknown *Contributors*: United States federal government ( converted it from JPEG to PNG and retouched it)

**File:Battle of Gettysburg, by Currier and Ives.png** *Source*: http://en.wikipedia.org/w/index.php?title=File:Battle_of_Gettysburg,_by_Currier_and_Ives.png *License*: unknown *Contributors*: Beyond silence, Churchh, Jweiss11, Kyro, Morgan Riley, Mtsmallwood, Möchtegern, PhilFree, Shizhao, Thuresson, UpstateNYer, Veikk0.ma, 18 anonymous edits

**Image:Ellis island 1902.jpg** *Source*: http://en.wikipedia.org/w/index.php?title=File:Ellis_island_1902.jpg *License*: unknown *Contributors*: AnRo0002, Aude, Infrogmation, Kjetil r, Philpill691, UpstateNYer, 9 anonymous edits

**Image:Into the Jaws of Death 23-0455M edit.jpg** *Source*: http://en.wikipedia.org/w/index.php?title=File:Into_the_Jaws_of_Death_23-0455M_edit.jpg *License*: unknown *Contributors*: Alvesgaspar, Apalsola, Armbrust, Berrucomons, Bjankuloski06en, Catsmeat, Christoph Braun, Esemono, Mogelzahn, Moroboshi, Nnemo, Noritaka666, Shizhao, Svajcr, Thierry Caro, Verdy p, Viloris

**Image:Abernathy Children on front line leading the SELMA TO MONTGOMERY MARCH for the RIGHT TO VOTE.JPG** *Source*: http://en.wikipedia.org/w/index.php?title=File:Abernathy_Children_on_front_line_leading_the_SELMA_TO_MONTGOMERY_MARCH_for_the_RIGHT_TO_VOTE.JPG *License*: unknown *Contributors*: Abernathy Family

**Image:DF-ST-91-01394.JPEG** *Source*: http://en.wikipedia.org/w/index.php?title=File:DF-ST-91-01394.JPEG *License*: unknown *Contributors*: DoD photo by: STAFF SERGEANT F. LEE COCKRAN

**File:UA Flight 175 hits WTC south tower 9-11 edit.jpeg** *Source*: http://en.wikipedia.org/w/index.php?title=File:UA_Flight_175_hits_WTC_south_tower_9-11_edit.jpeg *License*: unknown *Contributors*: User:UpstateNYer

**File:USA-satellite.jpg** *Source*: http://en.wikipedia.org/w/index.php?title=File:USA-satellite.jpg *License*: unknown *Contributors*: Original uploader was E Pluribus Anthony at en.wikipedia

**File:Bald Eagle Portrait.jpg** *Source*: http://en.wikipedia.org/w/index.php?title=File:Bald_Eagle_Portrait.jpg *License*: unknown *Contributors*: User:Saffron Blaze

**File:Census-2000-Data-Top-US-Ancestries-by-County.svg** *Source*: http://en.wikipedia.org/w/index.php?title=File:Census-2000-Data-Top-US-Ancestries-by-County.svg *License*: unknown *Contributors*: ANGELUS, Che829, Conscious, Davepape, Editor at Large, EurekaLott, Gpvos, JaumeBG, Jmabel, Kintetsubuffalo, MacedonianBoy, Man vyi, Mikael Häggström, Myself488, Nagy, Ras67, Tryphon, UpstateNYer, ZooFari, 17 anonymous edits

**File:Top of Rock Cropped.jpg** *Source*: http://en.wikipedia.org/w/index.php?title=File:Top_of_Rock_Cropped.jpg *License*: unknown *Contributors*: User:Dschwen

**File:LA Skyline Mountains2.jpg** *Source*: http://en.wikipedia.org/w/index.php?title=File:LA_Skyline_Mountains2.jpg *License*: unknown *Contributors*: User:Nserrano

**Image:Skyline_from_Millennium_Park.jpg** *Source*: http://en.wikipedia.org/w/index.php?title=File:Skyline_from_Millennium_Park.jpg *License*: unknown *Contributors*: User:Peterfitzgerald

**File:Religions of the United States.png** *Source*: http://en.wikipedia.org/w/index.php?title=File:Religions_of_the_United_States.png *License*: unknown *Contributors*: User:HaireDunya

**Image:Capitol Building Full View.jpg** *Source*: http://en.wikipedia.org/w/index.php?title=File:Capitol_Building_Full_View.jpg *License*: unknown *Contributors*: Noclip

**Image:WhiteHouseSouthFacade.JPG** *Source*: http://en.wikipedia.org/w/index.php?title=File:WhiteHouseSouthFacade.JPG *License*: unknown *Contributors*: Darwinius, Darxus, H005, HiraV, Infrogmation, J 1982, Jatkins, JenVan, Lycaon, Pi.1415926535, Redtigerxyz, TCY, UpstateNYer, 2 anonymous edits

**Image:USSupremeCourtWestFacade.JPG** *Source*: http://en.wikipedia.org/w/index.php?title=File:USSupremeCourtWestFacade.JPG *License*: unknown *Contributors*: User:UpstateNYer

**File:Political System of the United States.svg** *Source*: http://en.wikipedia.org/w/index.php?title=File:Political_System_of_the_United_States.svg *License*: unknown *Contributors*: User:111Alleskönner

**Image:Map of USA with state names 2.svg** *Source*: http://en.wikipedia.org/w/index.php?title=File:Map_of_USA_with_state_names_2.svg *License*: unknown *Contributors*: User:Andrew c

**File:Obama meets with Congressional Leadership July 2011.jpg** *Source*: http://en.wikipedia.org/w/index.php?title=File:Obama_meets_with_Congressional_Leadership_July_2011.jpg *License*: unknown *Contributors*: Antony-22

**File:Hague Clinton May 14 2010 Crop.jpeg** *Source*: http://en.wikipedia.org/w/index.php?title=File:Hague_Clinton_May_14_2010_Crop.jpeg *License*: unknown *Contributors*: User:UpstateNYer

**File:FederalDebt1940to2012.svg** *Source*: http://en.wikipedia.org/w/index.php?title=File:FederalDebt1940to2012.svg *License*: unknown *Contributors*: Mikhail Ryazanov

**File:US Navy 060618-N-8492C-212 An Air Force B-2 bomber along with other aircrafts from the Air Force, Navy and Marine Corps fly over the Kitty Hawk, Ronald Reagan and Abraham Lincoln Carrier Strike groups.jpg** *Source*: http://en.wikipedia.org/w/index.php?title=File:US_Navy_060618-N-8492C-212_An_Air_Force_B-2_bomber_along_with_other_aircrafts_from_the_Air_Force,_Navy_and_Marine_Corps_fly_over_the_Kitty *License*: unknown *Contributors*: Benchill, Docu, FAEP, FoolsWar, 1 anonymous edits

**File:NYPD impala.JPG** *Source*: http://en.wikipedia.org/w/index.php?title=File:NYPD_impala.JPG *License*: unknown *Contributors*: User:Pablo029

**File:Photos NewYork1 032.jpg** *Source*: http://en.wikipedia.org/w/index.php?title=File:Photos_NewYork1_032.jpg *License*: unknown *Contributors*: Basilicofresco, Beyond silence, Loranchet, Quasipalm, Urban, 1 anonymous edits

**File:Median US household income.png** *Source*: http://en.wikipedia.org/w/index.php?title=File:Median_US_household_income.png *License*: unknown *Contributors*: CLDunlap

**File:South San Jose (crop).jpg** *Source*: http://en.wikipedia.org/w/index.php?title=File:South_San_Jose_(crop).jpg *License*: unknown *Contributors*: User:Seano1

**File:Map of current Interstates.svg** *Source*: http://en.wikipedia.org/w/index.php?title=File:Map_of_current_Interstates.svg *License*: unknown *Contributors*: User:SPUI

**File:Neil Armstrong in suit.jpg** *Source*: http://en.wikipedia.org/w/index.php?title=File:Neil_Armstrong_in_suit.jpg *License*: unknown *Contributors*: User:W.wolny

**File:University-of-Virginia-Rotunda.jpg** *Source*: http://en.wikipedia.org/w/index.php?title=File:University-of-Virginia-Rotunda.jpg *License*: unknown *Contributors*: gargola87 on Flickr

**File:Aerial of Texas Medical Center with Downtown Houston in the background.jpg** *Source*: http://en.wikipedia.org/w/index.php?title=File:Aerial_of_Texas_Medical_Center_with_Downtown_Houston_in_the_background.jpg *License*: unknown *Contributors*: Agsftw, Hequals2henry

**File:Liberty-statue-from-below.jpg** *Source*: http://en.wikipedia.org/w/index.php?title=File:Liberty-statue-from-below.jpg *License*: unknown *Contributors*: User:Tysto

**File:Mark Twain, Brady-Handy photo portrait, Feb 7, 1871, cropped.jpg** *Source*: http://en.wikipedia.org/w/index.php?title=File:Mark_Twain,_Brady-Handy_photo_portrait,_Feb_7,_1871,_cropped.jpg *License*: unknown *Contributors*: Mathew Brady

**File:Times Square 1-2.JPG** *Source*: http://en.wikipedia.org/w/index.php?title=File:Times_Square_1-2.JPG *License*: unknown *Contributors*: Beyond My Ken, Gryffindor, Infrogmation, Jatkins, Jim.henderson, Julia W, LongLiveRock, Sdrtirs, Takabeg, UpstateNYer, 1 anonymous edits

**File:Motherhood and apple pie.jpg** *Source*: http://en.wikipedia.org/w/index.php?title=File:Motherhood_and_apple_pie.jpg *License*: unknown *Contributors*: Scott Bauer, USDA ARS

**File:Michael Phelps 2009.jpg** *Source*: http://en.wikipedia.org/w/index.php?title=File:Michael_Phelps_2009.jpg *License*: unknown *Contributors*: Marylandstater

**File:Office-book.svg** *Source*: http://en.wikipedia.org/w/index.php?title=File:Office-book.svg *License*: unknown *Contributors*: Chris Down/Tango project

**File:Gnome-globe.svg** *Source*: http://en.wikipedia.org/w/index.php?title=File:Gnome-globe.svg *License*: unknown *Contributors*: David Vignoni

**File:Pennsylvania-counties-map.gif** *Source*: http://en.wikipedia.org/w/index.php?title=File:Pennsylvania-counties-map.gif *License*: unknown *Contributors*: AnonMoos, Camildo, JosN, Padraic, 2 anonymous edits

**File:National-atlas-pennsylvania.png** *Source*: http://en.wikipedia.org/w/index.php?title=File:National-atlas-pennsylvania.png *License*: unknown *Contributors*: Shauni

**File:Edward Hicks - Penn's Treaty.jpeg** *Source*: http://en.wikipedia.org/w/index.php?title=File:Edward_Hicks_-_Penn's_Treaty.jpeg *License*: unknown *Contributors*: Cobalty, Man vyi, Uyvsdi, Zolo, 1 anonymous edits

**File:John Dickinson portrait.jpg** *Source*: http://en.wikipedia.org/w/index.php?title=File:John_Dickinson_portrait.jpg *License*: unknown *Contributors*: User:PKM

**File:PhiladelphiaPresidentsHouse.jpg** *Source*: http://en.wikipedia.org/w/index.php?title=File:PhiladelphiaPresidentsHouse.jpg *License*: unknown *Contributors*: Aavindraa, BoringHistoryGuy, 1 anonymous edits

**File:HillsCapitol.jpg** *Source*: http://en.wikipedia.org/w/index.php?title=File:HillsCapitol.jpg *License*: unknown *Contributors*: Original uploader was ClairSamoht at en.wikipedia

**File:Federal Emergency Relief Administration, FERA camp for unemployed women in Arcola, PA - NARA - 196583.tif** *Source*: http://en.wikipedia.org/w/index.php?title=File:Federal_Emergency_Relief_Administration,_FERA_camp_for_unemployed_women_in_Arcola,_PA_-_NARA_-_196583.tif *License*: unknown *Contributors*: Howcheng, Infrogmation

**Image:PennsylvaniaDemography.svg** *Source*: http://en.wikipedia.org/w/index.php?title=File:PennsylvaniaDemography.svg *License*: unknown *Contributors*: User:Quilokos

**File:Pennsylvania population map 1.png** *Source*: http://en.wikipedia.org/w/index.php?title=File:Pennsylvania_population_map_1.png *License*: unknown *Contributors*: User:JimIrwin

**File:Bethlehem Steel.jpg** *Source*: http://en.wikipedia.org/w/index.php?title=File:Bethlehem_Steel.jpg *License*: unknown *Contributors*: Original uploader was Jschnalzer at en.wikipedia

**Image:Crayola-64.jpg** *Source*: http://en.wikipedia.org/w/index.php?title=File:Crayola-64.jpg *License*: unknown *Contributors*: Dougie WII

**File:Philly skyline.jpg** *Source*: http://en.wikipedia.org/w/index.php?title=File:Philly_skyline.jpg *License*: unknown *Contributors*: Ed Yakovich , emy111 , flickr: i like the weekend , http://www.flickr.com/photos/10396190@N04/. Original uploader was Emy111 at en.wikipedia

**File:Pittsburgh WEO Night 1.jpg** *Source*: http://en.wikipedia.org/w/index.php?title=File:Pittsburgh_WEO_Night_1.jpg *License*: unknown *Contributors*: User:ronjamin

**File:Allentown.jpg** *Source*: http://en.wikipedia.org/w/index.php?title=File:Allentown.jpg *License*: unknown *Contributors*: Paul Leiby

**File:Penn campus 2.jpg** *Source*: http://en.wikipedia.org/w/index.php?title=File:Penn_campus_2.jpg *License*: unknown *Contributors*: Bryan Y.W. Shin

**Image:Dorney Park Steel Force Thunderhawk.jpg** *Source*: http://en.wikipedia.org/w/index.php?title=File:Dorney_Park_Steel_Force_Thunderhawk.jpg *License*: unknown *Contributors*: Themeparkgc

**File:Baughman's Rock Overlook.ogv** *Source*: http://en.wikipedia.org/w/index.php?title=File:Baughman's_Rock_Overlook.ogv *License*: unknown *Contributors*: David Fulmer from Pittsburgh

**Image:Allegheny Mountain Tunnel.JPG** *Source*: http://en.wikipedia.org/w/index.php?title=File:Allegheny_Mountain_Tunnel.JPG *License*: unknown *Contributors*: Taken by Willkie Cirker. User Bcirker on en.wikipedia

**Image:2215 - Pennsylvania Turnpike Tunnel.JPG** *Source*: http://en.wikipedia.org/w/index.php?title=File:2215_-_Pennsylvania_Turnpike_Tunnel.JPG *License*: unknown *Contributors*: User:Thisisbossi

**File:Appalachian Throughway.jpg** *Source*: http://en.wikipedia.org/w/index.php?title=File:Appalachian_Throughway.jpg *License*: unknown *Contributors*: Nicholas

**Image:Cokepark allentown.jpg** *Source*: http://en.wikipedia.org/w/index.php?title=File:Cokepark_allentown.jpg *License*: unknown *Contributors*: User:Spikerogan

**Image:Pocono Victory Lane.JPG** *Source*: http://en.wikipedia.org/w/index.php?title=File:Pocono_Victory_Lane.JPG *License*: unknown *Contributors*: Morio, STBR

**File:RuffedGrouse23.jpg** *Source*: http://en.wikipedia.org/w/index.php?title=File:RuffedGrouse23.jpg *License*: unknown *Contributors*: Unknown/Uncredited - assumed USFWS employee.. Original uploader was Big iron at en.wikipedia

**File:USBrigNiagaraInPort.JPG** *Source*: http://en.wikipedia.org/w/index.php?title=File:USBrigNiagaraInPort.JPG *License*: unknown *Contributors*: User:Howcheng

**File:Downtown Aliquippa.jpg** *Source*: http://en.wikipedia.org/w/index.php?title=File:Downtown_Aliquippa.jpg *License*: unknown *Contributors*: User:Mvincec

**File:AllentownPA Skyline.jpg** *Source*: http://en.wikipedia.org/w/index.php?title=File:AllentownPA_Skyline.jpg *License*: unknown *Contributors*: Thomas Grim - twg1942

**File:Altoona Downtown from Brush Mountain.jpg** *Source*: http://en.wikipedia.org/w/index.php?title=File:Altoona_Downtown_from_Brush_Mountain.jpg *License*: unknown *Contributors*: User:Csoldner187

**File:Beaver Falls.jpg** *Source*: http://en.wikipedia.org/w/index.php?title=File:Beaver_Falls.jpg *License*: unknown *Contributors*: MrHope, Nyttend

**File:Bedford pitt street.jpg** *Source*: http://en.wikipedia.org/w/index.php?title=File:Bedford_pitt_street.jpg *License*: unknown *Contributors*: User:AudeVivere

**File:Bethlehem Pennsylvania downtown.jpg** *Source*: http://en.wikipedia.org/w/index.php?title=File:Bethlehem_Pennsylvania_downtown.jpg *License*: unknown *Contributors*: Tim Kiser ()

**File:Blairsville Pennsylvania Market Street 2010.jpg** *Source*: http://en.wikipedia.org/w/index.php?title=File:Blairsville_Pennsylvania_Market_Street_2010.jpg *License*: unknown *Contributors*: User:Canadian2006

**File:Butler PA skyline.jpg** *Source*: http://en.wikipedia.org/w/index.php?title=File:Butler_PA_skyline.jpg *License*: unknown *Contributors*: Mvincec. Original uploader was Mvincec at en.wikipedia

**File:Carlisle, Pennsylvania.jpg** *Source*: http://en.wikipedia.org/w/index.php?title=File:Carlisle,_Pennsylvania.jpg *License*: unknown *Contributors*: Doug Kerr

**File:Memorial Square Chambersburg.jpg** *Source*: http://en.wikipedia.org/w/index.php?title=File:Memorial_Square_Chambersburg.jpg *License*: unknown *Contributors*: Epstein.Mark, Euchiasmus, Jmabel, Vpuliva, 1 anonymous edits

**File:Clarion, Pennsylvania.jpg** *Source*: http://en.wikipedia.org/w/index.php?title=File:Clarion,_Pennsylvania.jpg *License*: unknown *Contributors*: Doug Kerr

**File:Clearfield County Courthouse Apr 10.JPG** *Source*: http://en.wikipedia.org/w/index.php?title=File:Clearfield_County_Courthouse_Apr_10.JPG *License*: unknown *Contributors*: Mackensen

**File:Corry, Pennsylvania Skyline.jpg** *Source*: http://en.wikipedia.org/w/index.php?title=File:Corry,_Pennsylvania_Skyline.jpg *License*: unknown *Contributors*: Nicholas

**File:Pottercourthousecoudersport.jpg** *Source*: http://en.wikipedia.org/w/index.php?title=File:Pottercourthousecoudersport.jpg *License*: unknown *Contributors*: Choess, Mackensen

**File:Easton Skyline.jpg** *Source*: http://en.wikipedia.org/w/index.php?title=File:Easton_Skyline.jpg *License*: unknown *Contributors*: Original uploader was Masesprint at en.wikipedia

**File:Cambria County Courthouse - Ebensburg, PA.jpg** *Source*: http://en.wikipedia.org/w/index.php?title=File:Cambria_County_Courthouse_-_Ebensburg,_PA.jpg *License*: unknown *Contributors*: User:Publichall

**File:edinboro.jpg** *Source*: http://en.wikipedia.org/w/index.php?title=File:Edinboro.jpg *License*: unknown *Contributors*: John of Reading, Taurus9395

**File:Eriesky2.jpg** *Source*: http://en.wikipedia.org/w/index.php?title=File:Eriesky2.jpg *License*: unknown *Contributors*: Original uploader was Pnoble805 at en.wikipedia

**File:Franklin, Pennsylvania.jpg** *Source*: http://en.wikipedia.org/w/index.php?title=File:Franklin,_Pennsylvania.jpg *License*: unknown *Contributors*: Doug Kerr

**File:DowntownGettysburgPA.jpg** *Source*: http://en.wikipedia.org/w/index.php?title=File:DowntownGettysburgPA.jpg *License*: unknown *Contributors*: Aurora30

**File:Greencastle PA Square.JPG** *Source*: http://en.wikipedia.org/w/index.php?title=File:Greencastle_PA_Square.JPG *License*: unknown *Contributors*: User:Smallbones

**File:Greensburgskyline.jpg** *Source*: http://en.wikipedia.org/w/index.php?title=File:Greensburgskyline.jpg *License*: unknown *Contributors*: User:Neurodistortion

**File:Hanover theatre PA.jpg** *Source*: http://en.wikipedia.org/w/index.php?title=File:Hanover_theatre_PA.jpg *License*: unknown *Contributors*: JOHN LLOYD

**File:Downtown hazleton pa.jpg** *Source*: http://en.wikipedia.org/w/index.php?title=File:Downtown_hazleton_pa.jpg *License*: unknown *Contributors*: Fenderbob, 2 anonymous edits

**File:Hershey, Pennsylvania.JPG** *Source*: http://en.wikipedia.org/w/index.php?title=File:Hershey,_Pennsylvania.JPG *License*: unknown *Contributors*: Original uploader was Bronayur at en.wikipedia

**File:2008 03 30 - Hollidaysburg - Blair St 2.JPG** *Source*: http://en.wikipedia.org/w/index.php?title=File:2008_03_30_-_Hollidaysburg_-_Blair_St_2.JPG *License*: unknown *Contributors*: User:Thisisbossi

**File:Indiana Theater Sign.jpg** *Source*: http://en.wikipedia.org/w/index.php?title=File:Indiana_Theater_Sign.jpg *License*: unknown *Contributors*: User:Scsmatt

**File:Jeannette-pennsylvania-downtown.jpg** *Source*: http://en.wikipedia.org/w/index.php?title=File:Jeannette-pennsylvania-downtown.jpg *License*: unknown *Contributors*: User:Canadian2006

**File:Johnstownview.jpg** *Source*: http://en.wikipedia.org/w/index.php?title=File:Johnstownview.jpg *License*: unknown *Contributors*: Auntof6, Jllm06

**File:Kittanning fromhill.jpg** *Source*: http://en.wikipedia.org/w/index.php?title=File:Kittanning_fromhill.jpg *License*: unknown *Contributors*: All Is One, Cngodles, 1 anonymous edits

**File:Lancaster Pennsylvania downtown.jpg** *Source*: http://en.wikipedia.org/w/index.php?title=File:Lancaster_Pennsylvania_downtown.jpg *License*: unknown *Contributors*: Randolph Carney

**File:Latrobe PA MainStreet.jpg** *Source*: http://en.wikipedia.org/w/index.php?title=File:Latrobe_PA_MainStreet.jpg *License*: unknown *Contributors*: User:Ohnoitsjamie

**File:LebanonBolognaDrop2008.jpg** *Source*: http://en.wikipedia.org/w/index.php?title=File:LebanonBolognaDrop2008.jpg *License*: unknown *Contributors*: Hooahman

**File:Lewisburg, Pennsylvania.jpg** *Source*: http://en.wikipedia.org/w/index.php?title=File:Lewisburg,_Pennsylvania.jpg *License*: unknown *Contributors*: Doug Kerr

**File:ltownsquare1.jpg** *Source*: http://en.wikipedia.org/w/index.php?title=File:Ltownsquare1.jpg *License*: unknown *Contributors*: User:KATMAAN

**File:Clinton County Pennsylvania Courthouse.JPG** *Source*: http://en.wikipedia.org/w/index.php?title=File:Clinton_County_Pennsylvania_Courthouse.JPG *License*: unknown *Contributors*: User:Ruhrfisch

**File:McKeesRocksPA.jpg** *Source*: http://en.wikipedia.org/w/index.php?title=File:McKeesRocksPA.jpg *License*: unknown *Contributors*:

**File:Meadville, Pennsylvania.jpg** *Source*: http://en.wikipedia.org/w/index.php?title=File:Meadville,_Pennsylvania.jpg *License*: unknown *Contributors*: Doug Kerr

**File:Monessen01.JPG** *Source*: http://en.wikipedia.org/w/index.php?title=File:Monessen01.JPG *License*: unknown *Contributors*: User:Mlouns

**File:Kennedy Square.jpg** *Source*: http://en.wikipedia.org/w/index.php?title=File:Kennedy_Square.jpg *License*: unknown *Contributors*: Nyttend

**File:New Hope and Ivyland.JPG** *Source*: http://en.wikipedia.org/w/index.php?title=File:New_Hope_and_Ivyland.JPG *License*: unknown *Contributors*: User:Jared Kofsky

**File:Palmyra Downtown.jpg** *Source*: http://en.wikipedia.org/w/index.php?title=File:Palmyra_Downtown.jpg *License*: unknown *Contributors*: Jayu from Harrisburg, PA, U.S.A.

**File:Philadelphia Panorama From Camden.JPG** *Source*: http://en.wikipedia.org/w/index.php?title=File:Philadelphia_Panorama_From_Camden.JPG *License*: unknown *Contributors*: User:Parent5446

**File:Phoenixville downtown.jpg** *Source*: http://en.wikipedia.org/w/index.php?title=File:Phoenixville_downtown.jpg *License*: unknown *Contributors*: User:Squelle

**File:Duquesne Incline from top.jpg** *Source*: http://en.wikipedia.org/w/index.php?title=File:Duquesne_Incline_from_top.jpg *License*: unknown *Contributors*: User:Plastikspork

**File:Pittston City.jpg** *Source*: http://en.wikipedia.org/w/index.php?title=File:Pittston_City.jpg *License*: unknown *Contributors*: User:vasiliymeshko

**File:Pottstown PA HighStreet.jpg** *Source*: http://en.wikipedia.org/w/index.php?title=File:Pottstown_PA_HighStreet.jpg *License*: unknown *Contributors*: Willjay

**File:Pottsville.jpg** *Source*: http://en.wikipedia.org/w/index.php?title=File:Pottsville.jpg *License*: unknown *Contributors*: Twinmama (talk). Original text: by Angela Henderson.

**File:Groundhogday2005.jpg** *Source*: http://en.wikipedia.org/w/index.php?title=File:Groundhogday2005.jpg *License*: unknown *Contributors*: Evrik, FlickreviewR, GrapedApe, Guety, Quadell, 1 anonymous edits

**File:Reading, Pennsylvani skyline.jpg** *Source*: http://en.wikipedia.org/w/index.php?title=File:Reading,_Pennsylvani_skyline.jpg *License*: unknown *Contributors*: Jack

**File:Renovo, Pennsylvania 14 Street.jpg** *Source*: http://en.wikipedia.org/w/index.php?title=File:Renovo,_Pennsylvania_14_Street.jpg *License*: unknown *Contributors*: User:Ruhrfisch

**File:Ridgway PA from Country Club Apr 10.JPG** *Source*: http://en.wikipedia.org/w/index.php?title=File:Ridgway_PA_from_Country_Club_Apr_10.JPG *License*: unknown *Contributors*: Auntof6, GrapedApe, Leyo, Mjrmtg

**File:DowntownScranton2003.jpg** *Source*: http://en.wikipedia.org/w/index.php?title=File:DowntownScranton2003.jpg *License*: unknown *Contributors*: Don Liotta. Original uploader was Donaldliotta at en.wikipedia

**File:Selinsgrove, Pennsylvania.jpg** *Source*: http://en.wikipedia.org/w/index.php?title=File:Selinsgrove,_Pennsylvania.jpg *License*: unknown *Contributors*: Doug Kerr

**File:Uprising3.JPG** *Source*: http://en.wikipedia.org/w/index.php?title=File:Uprising3.JPG *License*: unknown *Contributors*: Hal Smith

**File:ThecornerSC.JPG** *Source*: http://en.wikipedia.org/w/index.php?title=File:ThecornerSC.JPG *License*: unknown *Contributors*: Rafaelgarcia, Thisisbossi, Thuresson

**File:Sunbury, Pennsylvania.jpg** *Source*: http://en.wikipedia.org/w/index.php?title=File:Sunbury,_Pennsylvania.jpg *License*: unknown *Contributors*: Doug Kerr

**File:Tyrone Borough Historic District Apr 10.JPG** *Source*: http://en.wikipedia.org/w/index.php?title=File:Tyrone_Borough_Historic_District_Apr_10.JPG *License*: unknown *Contributors*: Auntof6, DanTD, Leyo

**File:Uniontown PA.jpg** *Source*: http://en.wikipedia.org/w/index.php?title=File:Uniontown_PA.jpg *License*: unknown *Contributors*: Guswrites, 2 anonymous edits

**File:Warren Pennsylvania.jpg** *Source*: http://en.wikipedia.org/w/index.php?title=File:Warren_Pennsylvania.jpg *License*: unknown *Contributors*: Nicholas from Pennsylvania, USA

**File:WashingtonCountyCourthouse.JPG** *Source*: http://en.wikipedia.org/w/index.php?title=File:WashingtonCountyCourthouse.JPG *License*: unknown *Contributors*: GrapedApe, Nyttend

**File:Waynesboro, Pennsylvania.JPG** *Source*: http://en.wikipedia.org/w/index.php?title=File:Waynesboro,_Pennsylvania.JPG *License*: unknown *Contributors*: User:Dincher

**File:Wilkes Barre Downtown.jpg** *Source*: http://en.wikipedia.org/w/index.php?title=File:Wilkes_Barre_Downtown.jpg *License*: unknown *Contributors*: User:vasiliymeshko

**File:City Hall Williamsport Pennsylvania.JPG** *Source*: http://en.wikipedia.org/w/index.php?title=File:City_Hall_Williamsport_Pennsylvania.JPG *License*: unknown *Contributors*: User:Ruhrfisch

**File:Golden Plow York PA.JPG** *Source*: http://en.wikipedia.org/w/index.php?title=File:Golden_Plow_York_PA.JPG *License*: unknown *Contributors*: User:Smallbones

**Image:LebanonBolognaDrop2008.jpg** *Source*: http://en.wikipedia.org/w/index.php?title=File:LebanonBolognaDrop2008.jpg *License*: unknown *Contributors*: Hooahman

**file:Census Bureau seal.svg** *Source*: http://en.wikipedia.org/w/index.php?title=File:Census_Bureau_seal.svg *License*: unknown *Contributors*: United States Bureau of the CensusThis vector version by Mysid

**file:U.S. Census Bureau logo post-2011.png** *Source*: http://en.wikipedia.org/w/index.php?title=File:U.S._Census_Bureau_logo_post-2011.png *License*: unknown *Contributors*: U.S. federal government

**Image:U.S. Census Bureau Regions.svg** *Source*: http://en.wikipedia.org/w/index.php?title=File:U.S._Census_Bureau_Regions.svg *License*: unknown *Contributors*: U.S. Department of Commerce Economics and Statistics Administration U.S. CENSUS BUREAU

**Image:New Regional Offices.png** *Source*: http://en.wikipedia.org/w/index.php?title=File:New_Regional_Offices.png *License*: unknown *Contributors*: US Bureau of the Census

**File:Census Bureau headquarters, Suitland, Maryland, 2007.jpg** *Source*: http://en.wikipedia.org/w/index.php?title=File:Census_Bureau_headquarters,_Suitland,_Maryland,_2007.jpg *License*: unknown *Contributors*: United States Census Bureau

**File:Census Regions and Division of the United States.svg** *Source*: http://en.wikipedia.org/w/index.php?title=File:Census_Regions_and_Division_of_the_United_States.svg *License*: unknown *Contributors*: US Census Bureau

**File:This is a card puncher, an integral part of the tabulation system used by the United States Census Bureau to compile... - NARA - 513295.tif** *Source*: http://en.wikipedia.org/w/index.php?title=File:This_is_a_card_puncher,_an_integral_part_of_the_tabulation_system_used_by_the_United_States_Census_Bureau_to_compile..._-_NARA_-_513295.tif *License*: unknown *Contributors*: ArnoldReinhold, Infrogmation, Mattes, Umnik, Yann

**File:Volkstelling 1925 Census.jpg** *Source*: http://en.wikipedia.org/w/index.php?title=File:Volkstelling_1925_Census.jpg *License*: unknown *Contributors*: Aiko, Classical geographer, Frank C. Müller, JPS68, Vincent Steenberg, Vluebben

**Image:Countries by population density.svg** *Source*: http://en.wikipedia.org/w/index.php?title=File:Countries_by_population_density.svg *License*: unknown *Contributors*: Miguel Contreras, Guatemala

**File:World_population_density_1994.png** *Source*: http://en.wikipedia.org/w/index.php?title=File:World_population_density_1994.png *License*: unknown *Contributors*: User:QuartierLatin1968

**Image:Population density with key.png** *Source*: http://en.wikipedia.org/w/index.php?title=File:Population_density_with_key.png *License*: unknown *Contributors*: User:Keenan Pepper, User:SG

**File:Aavikko.png** *Source*: http://en.wikipedia.org/w/index.php?title=File:Aavikko.png *License*: unknown *Contributors*: User Vzb83 on fi.wikipedia

**Image:Monaco by night.JPG** *Source*: http://en.wikipedia.org/w/index.php?title=File:Monaco_by_night.JPG *License*: unknown *Contributors*: User:Ewa Dryjanska

**File:TariatLandscape.jpg** *Source*: http://en.wikipedia.org/w/index.php?title=File:TariatLandscape.jpg *License*: unknown *Contributors*: User:Doron

**File:EU Pop2008 1024.PNG** *Source*: http://en.wikipedia.org/w/index.php?title=File:EU_Pop2008_1024.PNG *License*: unknown *Contributors*: User:Vinny Burgoo

**File:Flag of Singapore.svg** *Source*: http://en.wikipedia.org/w/index.php?title=File:Flag_of_Singapore.svg *License*: unknown *Contributors*: Various

**File:Flag of Hong Kong.svg** *Source*: http://en.wikipedia.org/w/index.php?title=File:Flag_of_Hong_Kong.svg *License*: unknown *Contributors*: Designed by

**File:Flag of Bahrain.svg** *Source*: http://en.wikipedia.org/w/index.php?title=File:Flag_of_Bahrain.svg *License*: unknown *Contributors*: User:SKopp, User:Zscout370

**File:Flag of Bangladesh.svg** *Source*: http://en.wikipedia.org/w/index.php?title=File:Flag_of_Bangladesh.svg *License*: unknown *Contributors*: User:SKopp

**File:Flag of the Republic of China.svg** *Source*: http://en.wikipedia.org/w/index.php?title=File:Flag_of_the_Republic_of_China.svg *License*: unknown *Contributors*: 555, Abner1069, Bestalex, Bigmorr, Bjankuloski06en, Denelson83, Ed veg, Gzdavidwong, Herbythyme, Isletakee, Kakoui, Kallerna, Kibinsky, MAXXX-309, Mattes, Mizunoryu, Neq00, Nickpo, Nightstallion, Odder, Pymouss, R.O.C, Reisio, Reuvenk, Rkt2312, Rocket000, Runningfridgesrule, Samwingkit, Sasha Krotov, Shizhao, SiBr4, Tabasco, Theo10011, Vzb83, Wrightbus, ZooFari, Zscout370, 75 anonymous edits

**File:Flag of Mauritius.svg** *Source*: http://en.wikipedia.org/w/index.php?title=File:Flag_of_Mauritius.svg *License*: unknown *Contributors*: User:Zscout370

**File:Flag of South Korea.svg** *Source*: http://en.wikipedia.org/w/index.php?title=File:Flag_of_South_Korea.svg *License*: unknown *Contributors*: Various

**File:Flag of Rwanda.svg** *Source*: http://en.wikipedia.org/w/index.php?title=File:Flag_of_Rwanda.svg *License*: unknown *Contributors*: .

**File:Flag of the Netherlands.svg** *Source*: http://en.wikipedia.org/w/index.php?title=File:Flag_of_the_Netherlands.svg *License*: unknown *Contributors*: User:Zscout370

**File:Flag of Lebanon.svg** *Source*: http://en.wikipedia.org/w/index.php?title=File:Flag_of_Lebanon.svg *License*: unknown *Contributors*: Traced based on the CIA World Factbook with some modification done to the colours based on information at Vexilla mundi.

**File:Flag of India.svg** *Source*: http://en.wikipedia.org/w/index.php?title=File:Flag_of_India.svg *License*: unknown *Contributors*: Anomie, Mifter

**File:Flag of Belgium (civil).svg** *Source*: http://en.wikipedia.org/w/index.php?title=File:Flag_of_Belgium_(civil).svg *License*: unknown *Contributors*: Bean49, Cathy Richards, David Descamps, Dbenbenn, Denelson83, Evanc0912, Fry1989, Gabriel trzy, Howcome, IvanOS, Mimich, Ms2ger, Nightstallion, Oreo Priest, Ricordisamoa, Rocket000, Rodejong, SiBr4, Sir Iain, ThomasPusch, Warddr, Zscout370, 6 anonymous edits

**File:Flag of Japan.svg** *Source*: http://en.wikipedia.org/w/index.php?title=File:Flag_of_Japan.svg *License*: unknown *Contributors*: Anomie

**File:Flag of Sri Lanka.svg** *Source*: http://en.wikipedia.org/w/index.php?title=File:Flag_of_Sri_Lanka.svg *License*: unknown *Contributors*: Zscout370

**Image:Time zone chicago.jpg** *Source*: http://en.wikipedia.org/w/index.php?title=File:Time_zone_chicago.jpg *License*: unknown *Contributors*: User:JoeSmack

**File:Standard time zones of the world.png** *Source*: http://en.wikipedia.org/w/index.php?title=File:Standard_time_zones_of_the_world.png *License*: unknown *Contributors*: User:TimeZonesBoy

**File:UTC hue4map X world Robinson.png** *Source*: http://en.wikipedia.org/w/index.php?title=File:UTC_hue4map_X_world_Robinson.png *License*: unknown *Contributors*: User:Phoenix B 1of3

**Image:Tzdiff-Europe-summer.png** *Source*: http://en.wikipedia.org/w/index.php?title=File:Tzdiff-Europe-summer.png *License*: unknown *Contributors*: Jmabel, Naive cynic

**Image:DaylightSaving-World-Subdivisions.png** *Source*: http://en.wikipedia.org/w/index.php?title=File:DaylightSaving-World-Subdivisions.png *License*: unknown *Contributors*: Paul Eggert

**File:Coventry Time Zone Clock.jpg** *Source*: http://en.wikipedia.org/w/index.php?title=File:Coventry_Time_Zone_Clock.jpg *License*: unknown *Contributors*: User:Cmglee

**File:Percentage population living on less than $1.25 per day 2009.svg** *Source*: http://en.wikipedia.org/w/index.php?title=File:Percentage_population_living_on_less_than_$1.25_per_day_2009.svg *License*: unknown *Contributors*: Tony0106

**File:Percent poverty world map.png** *Source*: http://en.wikipedia.org/w/index.php?title=File:Percent_poverty_world_map.png *License*: unknown *Contributors*: User:Sbw01f

MIX
Papier aus verantwortungsvollen Quellen
Paper from responsible sources
FSC® C105338

Printed by Books on Demand GmbH, Norderstedt / Germany